Analysis and Design of Machine Learning Techniques

Patrick Stalph

Analysis and Design of Machine Learning Techniques

Evolutionary Solutions for Regression, Prediction, and Control Problems

 Springer Vieweg

Patrick Stalph
Tübingen, Germany

PhD Thesis, University of Tübingen, 2013

ISBN 978-3-658-04936-2   ISBN 978-3-658-04937-9 (eBook)
DOI 10.1007/978-3-658-04937-9

The Deutsche Nationalbibliothek lists this publication in the Deutsche Nationalbibliografie; detailed bibliographic data are available in the Internet at http://dnb.d-nb.de.

Library of Congress Control Number: 2014931388

Springer Vieweg
© Springer Fachmedien Wiesbaden 2014

Printed on acid-free paper

Springer Vieweg is a brand of Springer DE.
Springer DE is part of Springer Science+Business Media.
www.springer-vieweg.de

Acknowledgments

First, I'd like to thank my supervisor, Martin Butz, for his support in general and, particularly, his constructive criticism, when it came to scientific writing. Furthermore, I want to thank the members of the department of cognitive modelling, formerly called COBOSLAB, for all those inspiring discussions.

I also want to thank Moritz Strübe, David Hock, Andreas Alin, Stephan Ehrenfeld, and Jan Kneissler for helpful reviews. Most importantly, I'm grateful for my wife being so patient with me.

Patrick Stalph

Abstract

Manipulating or grasping objects seems like a trivial task for humans, as these are motor skills of everyday life. However, motor skills are not easy to learn – babies require several month to develop proper grasping skills. Learning motor skills is also an active research topic in robotics. However, most solutions are optimized for industrial applications and, thus, few are plausible explanations for learning in the human brain.

The fundamental challenge, that motivates this research work, originates from the cognitive science: *How do humans learn their motor skills?* This work makes a connection between robotics and cognitive sciences by analyzing *motor skill learning* in well defined, analytically tractable scenarios using algorithmic implementations that could be found in human brain structures – at least to some extent. However, the work is on the technical side of the two research fields and oriented towards robotics.

The first contribution is an analysis of algorithms that are *suitable* for motor skill learning and *plausible* from a biological viewpoint. The identification of elementary features inherent to those algorithms allows other researchers to develop powerful, yet plausible learning algorithms for similar scenarios. The algorithm of choice, a learning classifier system – originally envisioned as a *cognitive system* –, is a genetic-based machine learning method that combines reinforcement learning and genetic algorithms. Two alternative algorithms are also considered.

The second part of this thesis is a scalability analysis of the learning classifier system and shows the limitations on the one hand, but on the other hand highlights ways to improve the scaling on certain problems. Nonetheless, high-dimensional problem may still require an unacceptable amount of training time. Therefore a new, informed search operator is developed to guide evolution through high-dimensional search spaces. Both the need for *specialization* but also *generalization* capabilities are integrated and thus the learning time is reduced drastically on high-dimensional problems.

The third part of this work discusses the basics of robot control and its challenges. A complete robot control framework is developed for a simulated, anthropomorphic arm with seven degrees of freedom. The aforementioned learning algorithms are embedded in a cognitive-plausible fashion. Finally, the framework is evaluated in a more realistic scenario with the anthropomorphic iCub robot, where stereo cameras are used for visual servoing. Convincing results confirm the applicability of the approach, while the biological plausibility is discussed in retrospect.

Contents

List of Figures

List of Algorithms

Glossary

ANN	Artificial Neural Network. 21, 22, 27, 138
control space	The space of control commands, e.g. torques or joint velocities. 3, 89, 90, 92, 93, 96, 97, 99, 102, 104–109, 115–120, 122, 132
DoF	Degrees of Freedom. 3, 7, 8, 87, 89–92, 101, 102, 104, 107, 108, 110, 111, 113, 115, 117, 118, 122, 139
FA	Function Approximation. 6, 11–13, 21, 26–28, 41, 77, 101, 102, 106, 107, 131, 137, 138, 142
forward kinematics	A mapping from joint configuration to task space location. XVII, 88, 89, 92, 101, 102, 106, 107, 110–112, 114, 115, 117, 118, 122, 125, 132, 140
GA	Genetic Algorithm. 26, 29, 41, 42, 45, 46, 48–50, 52, 58–64, 68, 69, 72, 73, 77–82, 101, 104, 115, 116
GPR	Gaussian Process Regression. 19, 20, 23, 24, 27, 30, 31
iCub	Icub is a humanoid robot. 1, 4, 8, 125, 126, 128–133, 140, 143
inverse kinematics	Inversion of the forward kinematics. 88, 92, 96, 97, 130, 142
Jacobian	A joint configuration dependent matrix that specifies the effect of joint angle changes on task space state. 89–93, 96, 97, 99–106, 108, 111–113, 130, 131
kernel	is a positive semi-definite matrix that defines a distance metric. 19, 21–38, 41, 43, 46, 47, 49, 51, 53, 63, 78, 82, 103, 120, 138

subsumption   Accurate receptive fields may *subsume* other more
              specific receptive fields to foster generalization. 50,
              52, 53, 58
SVD   Singular Value Decomposition. 75, 92, 93, 96, 97, 99,
              141

task space   The space where tasks are defined, e.g. Cartesian
              location of the end effector. 3, 5, 89–94, 96–99, 102,
              104–112, 115, 118–122, 125, 126, 129–132, 139, 140,
              142, 143

XCS   Accuracy based Learning Classifier System intro-
              duced by S. Wilson, sometimes called eXtended
              Learning Classifier System. 41, 50, 52, 57, 58
XCSF   Function approximation mode of XCS. 8, 26, 28–30,
              37, 39, 41, 42, 47, 48, 50, 52, 53, 57, 60–65, 67–69,
              71–74, 77–83, 101–106, 110, 111, 114–117, 119, 121–
              123, 125, 134, 138–141, 143

1 Introduction and Motivation

From a computational perspective, it is fascinating to see how fast humans or animals adapt to new, unforeseen situations. While an ordinary computer program can only handle foreseen situations it is applied to, humans are able to quickly estimate suitable behavior for new, unknown scenarios. A notable exception are Machine Learning (ML) algorithms that learn behavior based on some kind of optimization criterion. "Behavior of a computer program" may be a simple yes or no decision, or a rational choice of a robot's next action.

The first key in learning is to *memorize*, that is, to act optimal or rational in known situations. The next idea is to *generalize*, that is, to act well in unknown, but related situations. While ML is applied in various domains, including pattern recognition, language processing, data mining, computer games, and many more, the key question in this work is *how motor skills can be learned*.

Motor skills include pointing to objects, reaching movements, grasping, object manipulation, and locomotion, e.g. walking. The reader may now either think of robots or humans, maybe animals – and several different research disciplines come into mind. Learning robot motion is studied in Artificial Intelligence and Robotics. Human (or animal) learning, including motor skill learning, is studied in Neuroscience and Psychology. Finally, Cognitive Science is the interdisciplinary conjunction of those.

While the motivation and grounding research questions of this work originate from cognitive science, the focus is on the technical, roboticist side. Therefore the *cognitive plausibility* is a constraint throughout this work on ML algorithms that learn to control robot devices. Thus, the algorithms must adhere to some principles found in human brain structures and may therefore eventually serve as models for human learning. Furthermore, the robot devices in consideration are anthropomorphic ones, e.g. the humanoid iCub robot [69].

The following section introduces the major research question driving this work. Two more follow-up sections discuss the topic from the viewpoint of robotics – interested in fast moving, high precision robots – and also from a cognitive viewpoint based on the scientific interest in human brain structures that allow us to learn our motor skills so effectively.

1.1 How to Learn Motor Skills?

Humans or animals use their body to manipulate their surroundings or move themselves through this world. *Learning* motor skills requires knowledge of the relation between muscle activation and consequent movement. The term *motor skills* should not be confused with *motor primitives*, such as "grab", "reach", or "point to". Instead, motor skills refer to the knowledge of the effects of muscle activation – in order to achieve certain movements or motor primitives. Similarly, a robot controller must know the effects of motor commands.

Before we delve into the topic of learning, a few concepts have to be explained briefly. In a simplified view, human and robotic limbs can both be described as kinematic chains, that is, a number of rigid bodies connected by rotational joints as shown in Figure 1.1. While a two-joint robot arm is shown here, this work applies to other robot devices or human kinematic chains (e.g. the arm, or a leg) as well.

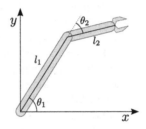

Figure 1.1: A simple planar arm with two links of lengths l_1 and l_2. The "shoulder" joint rotates about the origin with angle θ_1, while θ_2 describes the "elbow" angle.

First, a simple question has to be answered: *Why* would a robot (human, animal) move? The question for *motivation* is an important one in several aspects. First, without some sort of motivation, no movement is necessary. Trees are quite happy without moving around; there is no need for locomotion. However, animals hunt for pray or must flee from predators to survive. Humans walk to the food court for a tasty sushi. In the end it is Darwin's idea of survival that fuels the motivation of many actions. The same holds for robots: Why build a robot without a task? However, evolution is here represented by engineers that develop new robots, while the society decides which ones are needed or wanted. Again it is all about the "survival of the fittest".

Once there is motivation to reach a certain physical (not mental) goal, motor skills can be used to fulfill the task. For the sake of simplicity let us consider the robot arm in Figure 1.1 and a reaching task, that is, moving the grabber to a certain target location. Mathematically speaking, the task is to minimize the distance from grabber to target. This essentially defines the task space – for the simple planar arm this is a two-dimensional space of (x, y) locations. Furthermore, a task space can impose *kinematic redundancy* as shown in Figure 1.2, where a single target location in the 2D task space can be reached with two different joint configurations. Other tasks may impose a different task space: When a human is reaching for a cup of coffee, the orientation of the hand is important as well. The available control possibilities essentially define the Degrees of Freedom (DoF); e.g. a human arm has seven joints not including the fingers. The dimension of the task space is not necessarily equal to the number of DoF. When the DoF exceed the task space dimension, the kinematic chain is said to be *redundant* with respect to the task and this freedom can be used for secondary constraints.

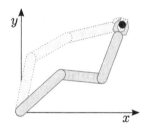

Figure 1.2: Often multiple joint configurations correspond to the same particular end effector location – also called kinematic redundancy.

Given a kinematic chain and a task, the next facet is *control*. Put differently, what control commands are required to fulfill the task at hand? Available control commands are specific to the kinematic chain and the space of control commands is the so called control space; be it muscle activation, joint velocity, or something completely different. In any case, the control commands uniquely specify the behavior of a kinematic device (not speaking about noise, yet).

Finally, *learning* comes into play. When the controller, e.g. a human, animal, or a silicium part of the robot, is able to learn an accurate mapping from control space to resulting movements in task space, this mapping can be used to solve given tasks.

To sum up, the motor skills are defined by a kinematic chain. Some sort of motivation (e.g. hunger or a robot's job) specifies a goal, which in turn defines the task space. To accomplish the given task, a *control strategy* based on a mapping from control space to task space is needed. Finally, such a mapping can be *learned* as a baby learns to control its hand and arm for grasping.

(a) real robot   (b) simulation

Figure 1.3: The iCub robot is available as physical entity and in simulation as well. Permission to reproduce image (a) was granted by the Robotcub Consortium, www. robotcub.org.

1.2 The Robotics Viewpoint

Roboticists might argue that robot kinematics are well studied and high precision solutions are available for realtime environments. Learning is of lesser interest, as models of industrial robots are known and the respective actuators work with high accuracy. Fast, analytical, stable solutions are available for typical kinematic chains. In short: Often learning is not required and hard-coded models are preferred for their simplicity and accuracy.

Learning becomes relevant when considering wear, breakage, calibration, or unpredictable external influences. For humanoid robots such as the iCub [69] shown in Figure 1.3, rather small servo motors are used and movements are not as accurate as those of heavy industrial robots. Thus, imperfect calibration is an issue that can be tackled with a learning approach that continuously adapts an underlying model.

In an industrial environment, often the same task has to be repeated over and over. Usually there is little need for flexible execution, but a static, predictable scheme is preferred. This also reduces the model complexity, as a small part of the full model is sufficient for a certain task.

Additionally, to be viable for a real robot, computational complexity of the learning approach is a critical part. However, learning an accurate kinematic model can be quite demanding and mobile robots have limited resources. The computational time of a control loop (sensing and computation of suitable motor commands) is crucial as it defines the granularity of control. Furthermore, accurate models may require a fair amount of memory. Thus, the computational complexity must be feasible for realtime control with respect to available hardware..

To sum up, the requirements for control of a real robot are accuracy, reliability, and eventually a suitable computational complexity when working with a mobile robot. A learning approach renders recalibration unnecessary. However, stability of the learned model must be guaranteed.

1.3 From the View of Cognitive Science

Humans cannot have a hard-coded, predefined, perfect model of their own body as the body changes over time [77] – a child is growing, an old mans bones ache. Studies have shown that adult mammalian brain structures can even adapt to extreme sensorimotor changes, e.g. to the loss of a limb [49, 50]. Clearly, there is some kind of learning mechanism and it is a continuous process throughout life.

However, learning is not required for everything. For example, suppose a human is reaching for a cup at the breakfast table. Obstacles may affect the required trajectory to grab the cup, but learning is not required for every obstacle anew. It is not reasonable to learn models for every type and location of obstacle (or constraints in general). Instead, the trained kinematic model should *know* about available kinematic redundancy (cf. Figure 1.2) that can be exploited on demand to solve constrained tasks.

However, many approaches resolve redundant alternatives during learning [94, 52, 48, 101, 39, 8] and consequently there is only a single joint configuration available for a certain task space location during control. Those models require less computational resources but, on the other hand, the controller has no movement alternatives for any given situation, even on highly redundant kinematic chains. This does not seem biologically plausible, even if one could argue that multiple such models can be learned for different constraints. Instead, kinematic redundancy should be learned [71, 15], which allows to flexibly use the full power of a given kinematic chain by resolving redundancy on the fly to achieve task-dependent optimal performance.

While computers work on a binary representation, that is, a single bit can be either zero or one, a brain is made of neurons that show highly complex behavior. First, the output of a neuron is a continuous value in contrast to a binary representation. Second, neurons fire impulses, eventually periodically with variable frequency instead of the fixed state of a bit. Third, communication in the brain is asynchronous, that is, neurons do not fire synchronously together, while typical computer simulations run a single (or few) synchronized processes. Consequently, a computer simulation can hardly model higher level brain functions in a *realistic* fashion [46]. However, the basic properties of neuronal activity can be captured in simple computational models and, thus, provide some explanatory power to eventually serve as a model of brain function.

On a higher level, human brain structures are partially organized in so called *population codes* [30, 68, 65, 1, 60, 76], that is, populations of neurons that encode a certain kind of space, e.g. a posture space. Furthermore, sensorimotor population codes have shown to represent *directional* encodings, that is, single cells correspond to a certain movement direction [31, 30, 1, 11]. Thus, using a directional population code further reduces the gap between computational models and real brain function.

To summarize, a biologically plausible model of motor skill learning should include knowledge about kinematic redundancy as humans (and animals) are good at flexibly moving under constraints, such as fixed joints, or moving around obstacles. Furthermore, it is desirable to have a neurally plausible model. In this regard, it has been shown that sensorimotor spaces are structured as population codes, that is, populations of neurons that represent a certain space.

1.4 Requirements for Biologically Plausible and Computationally Feasible Models

This section merges the requirements from the robotic and cognitive viewpoint to specify requirements for algorithms and methods in question.

Machine Learning (ML) is the keyword for *learning* approaches in computer science. Mammals cannot have a hard-coded model of their body readily available, as the body changes and, thus, learning of a body model is a key feature in this work.

Function Approximation (FA) is well suited to estimate an unknown kinematic model. A similar term from statistics is *regression*, which is slightly more concerned with variability and noise. FA is one way of ML, as the approximation can be seen as a learning process.

Online Learning and Control means that learning takes place *during* execution as opposed to separated batch-training and execution patterns, where large amounts of training data are fed to the learner, and, in turn, an executing controller uses the model without further learning. Put into biological terms, *learning* and *use* of motor skills are not separated in time, but are closely related.

Computational Complexity should remain in acceptable bounds such that the system is able to work online. On the one hand, this is a *runtime* restriction in the sense that a control loop cannot compute a motor command, say for one minute. Instead, a fast, highly frequent control loop assures smooth movements in dynamic environments. On the other hand, the complexity constraint restricts *memory* to some reasonable value. Excessive memory is neither available in the brain nor on a mobile robot.

Redundancy Preservation is desirable when a kinematic chain has redundant DoF that can be used to avoid obstacles or satisfy secondary constraints such as comfortable movements. Thus, redundant movement alternatives should be learned as it provides the flexibility needed to cope with unforeseen situations.

Neuronal Structures provide the model with biological plausibility and, thus, such a model could help to better understand how learning may take place in humans or animals. Since accurate computer simulations of realistic neurons are too demanding, a balance between realism and computational complexity has to be found.

Directional Population Codes have shown to encode sensorimotor mappings and, thus, provide another source of plausibility for a computational model of brain function. Mapping *changes* in joint configuration onto *changes* in hand location provides a simple model for directional motor control.

Autonomous Learning means knowledge gain through interaction with the environment without external guidance. This is another plausibility demand in that a human (animal) brain may come with some smart structure that was evolved over millions of years, but each individual must learn to control its body on its own.

In short, an online function approximation algorithm based on populations of neurons shall be applied to learn motor skills for flexible control. Given this list of key features and constraints, the following section outlines the scope and structure of the present work.

1.5 Scope and Structure

The focus of this thesis lies on computational, mathematical theory, and applicability in simulation, as a real robot was not available at the time of writing. Biological plausibility is a mandatory constraint to be satisfied as good as possible while maintaining a computationally feasible approach. Therefore the present work can be seen as a grounding research about the learning of motor skills from a computational and mathematical perspective. The remainder of this work can be separated into three parts: background knowledge, function approximation theory, and application to robot control.

Part I introduces general function approximation in Chapter 2. Algorithmic key features that satisfy the constraint of biological plausibility are identified. Chapter 3 describes these grounding elements in detail to allow for a deeper understanding of their inherent mathematical properties and identifies three suitable algorithms: Radial Basis Function Networks (RBFNs), the Locally Weighted Projection Regression (LWPR) algorithm, and the XCSF learning classifier system. As XCSF is the most promising candidate, its detailed algorithmic description is given in Chapter 4. Part II features a theoretical analysis of XCSF and its internal mechanisms in Chapter 5. Challenging problems and possible enhancements to overcome learning difficulties are proposed in Chapter 6.

Finally, all three algorithms are applied to robot control in Part III, where Chapter 7 introduces background information about robot kinematics in general. Next, the algorithms learn to control simulated, anthropomorphic arms of three to seven DoF in Chapter 8. Finally, a more realistic scenario is considered in Chapter 9, where a learning framework for the simulated iCub robot is built. Challenges such as noisy vision and the control of neck joints to follow targets lead to an interesting learning environment. The final Chapter 10 gives a summary and outlines possible extensions for future work. Concluding remarks complete the work.

Part I

Background

2 Introduction to Function Approximation and Regression

The first chapter of this thesis introduces Function Approximation (FA), also called *regression*, which is the basic feature required to learn sensorimotor mappings. A multitude of algorithm classes are introduced, including simple model fitting, interpolation, and advanced concepts such as Gaussian Processes and Artificial Neural Networks. The last section of this chapter discusses the applicability, but also questions the plausibility of such algorithms in the light of brain functionality.

2.1 Problem Statement

Continuous functions can be approximated with different methods depending on the type of function and the requirements to the resulting model. A function

$$f : X \rightarrow Y$$
$$f(x) = y$$

maps from $x \in X$ to $y \in Y$, where X is called the *input space* and Y is the *output space* of the function[1]. Generally, the data is available as samples (x, y), where the distribution of the inputs x and the function f are both unknown as illustrated in Figure 2.1. Eventually, the input space is multi-dimensional $X \subseteq \mathbb{R}^n$. In this case, n-dimensional column vectors $\boldsymbol{x} \in \boldsymbol{X}$ are written as bold symbols.

The task is to find a model $h(\boldsymbol{x})$ that explains the underlying data, that is, $h(\boldsymbol{x}) \approx y$ for all samples $(\boldsymbol{x}, y)$. Ideally the model equals the underlying function: $h(\boldsymbol{x}) = f(x)$. However, the function f can have some unknown properties at a subspace where no samples are available. Here, Occam's razor can be applied: Given two models that both accurately explain the data, the simpler model should be preferred. For example, data from a quadratic function should not be explained with a fourth order polynomial.

An alternative term for FA is regression, which does essentially the same thing (create some kind of model from given data), but focuses more on statistical

---

[1] In mathematics literature X may be called domain and Y the co-domain.

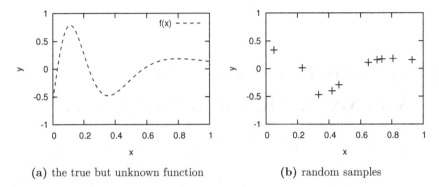

(a) the true but unknown function   (b) random samples

Figure 2.1: Illustration of a function approximation task. (a) The underlying function is usually unknown. (b) It can be challenging to accurately approximate a function surface from random samples. The data and its underlying function are specified in Appendix A.

properties such as expectation and variance. Usually regression approaches explicitly model noise or the variance of the prediction error in addition to the function surface. Input space X becomes the independent variable, where output space Y becomes the dependent variable. Throughout this work, the term *function approximation* is preferred, as noise poses an interesting challenge but need not be modeled explicitly and statistics are of lesser interest.

2.2 Measuring Quality

The goal of FA is to minimize the model (prediction, forecast) error. However, there are many ways to *measure* error that have only subtle differences. Given k samples (x, y) the quality of a model h can be assessed as follows. An intuitive measure is the Mean Absolute Error (MAE) computed as

$$\frac{1}{k} \sum_{j=1}^{k} |y_j - h(x_j)| \,, \tag{2.1}$$

which measures the average deviation from the true value.

The MAE does not account for the variance of the model error or, put differently, does not penalize outliers. Suppose model A accurately models 95% of the data but completely disagrees on the remaining five percent, while model B shows small errors over the full range of the data. They may well have the same MAE. When every data point has the same importance, model B may be preferable as it's errors have a lower variability. The amount of variability is taken into account by the following two measures.

Using the sum of *squared* errors instead of absolute errors yields the Mean Squared Error (MSE)

$$\frac{1}{k} \sum_{j=1}^{k} (y_j - h(x_j))^2 \, , \tag{2.2}$$

where error variance also affects the resulting value. However, the MSE is not as easily interpreted as the MAE, e.g. because units are squared.

Taking the square root yields the Root Mean Squared Error (RMSE)

$$\sqrt{\frac{1}{k} \sum_{j=1}^{k} (y_j - h(x_j))^2} \, , \tag{2.3}$$

which accounts for model deviation, error variance, but also more clearly relates to individual errors. When all errors have equal magnitude, then MAE and RMSE are equal as well; otherwise the RMSE is larger due to variability in the errors. When *accidental* outliers are known to be present, the RMSE may give a false impression of the model quality. Thus, the choice of quality measure depends on the actual application. Other quality measures exist, but those are usually tailored for a particular application. The MAE, MSE, and RMSE are the most common measures for FA.

2.3 Function Fitting or Parametric Regression

In the simplest form of FA the type of function is known and only a finite set of parameters has to be estimated. Put differently, the function is *fitted* to the data, e.g. a linear function or a polynomial. The extra assumption about the type of function not only improves, but also simplifies the approximation process. However, that assumption is a strong one to make and often the true underlying function is unknown. Thus, the chosen function type is a best guess and making wrong assumptions may result in poor quality models.

2.3.1 Linear Models with Ordinary Least Squares

Linear regression approximates the data with a linear function of two parameters $h(x) = \alpha + \beta x$. Intercept α and gradient β can be easily computed using the method of least squares [57, 29], where the squared error (thus, the RMSE as well) is minimized. Without particular assumptions on the distribution of noise (if there is any), the so called Ordinary Least Squares (OLS) method [53] can be applied to k data points:

$$\beta = \frac{\sum_{j=1}^{k}(x_j - \bar{x})(y_j - \bar{y})}{\sum_{j=1}^{k} x_j^2}, \quad \alpha = \bar{y} - \beta\bar{x} \tag{2.4}$$

where $\bar{x} = 1/k \sum_j x_j$ is the mean of the inputs and analogously $\bar{y}$ the mean of the function values. The above equation refers to one-dimensional inputs x.

The generalization to n dimensions is best described with a matrix formulation. Therefore, let $\boldsymbol{X}$ the $k \times n$ matrix of all inputs and $\boldsymbol{y}$ the vector of function values such that each row represents a sample $(\boldsymbol{x}_j, y_j)$. A linear mapping of all samples can be written as

$$\boldsymbol{X}\beta = \boldsymbol{y} \tag{2.5}$$

where the linear weights β are unknown. If the underlying data is truly linear, then there exists a solution to the above stated system of equations by inversion of the matrix $\boldsymbol{X}$ written as

$$\beta = \boldsymbol{X}^{-1}\boldsymbol{y}. \tag{2.6}$$

However, if the data is non-linear or noisy an approximation is required. The OLS estimator is calculated as

$$\beta = \left(\boldsymbol{X}^T\boldsymbol{X}\right)^{-1}\boldsymbol{X}^T\boldsymbol{y}, \tag{2.7}$$

where $\cdot^T$ is the transpose operator. The term $(\boldsymbol{X}^T\boldsymbol{X})^{-1}\boldsymbol{X}^T$ actually computes the so called Pseudoinverse matrix[2] which is the closest solution to a matrix inversion, if the regular inverse X^{-1} does not exist. The Pseudoinverse is equivalent to the regular inverse, if it exists.

The above formula assumes that the function values y have zero mean, that is, no intercept α is required. If this is not the case, the function values can be zero-centered by subtraction of the mean $y_i - \bar{y}$. Alternatively, a simple trick allows to include a non-zero intercept: The matrix $\boldsymbol{X}$ is extended with another column of ones, that is, $\boldsymbol{X}_{n+1} = 1$. The corresponding entry β_{n+1} then represents the intercept of the least squares hyperplane such that

$$h(x_1, \ldots, x_n) = \underbrace{\beta_1 x_1 + \ldots + \beta_n x_n}_{\text{linear model}} + \underbrace{\beta_{n+1}}_{\text{intercept}} \approx y. \tag{2.8}$$

---

[2]Details about computation of the Pseudoinverse, also known as Moore-Penrose matrix can be found in Section 7.3

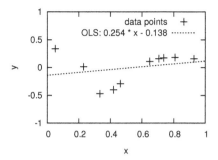

Figure 2.2: The linear OLS solution does not fit well to the underlying non-linear data. Furthermore, it is difficult to guess the true, underlying function type.

This simple OLS computation produces a linear model with minimal RMSE, even if the underlying data is *not* linear as exemplified in Figure 2.2. However, it requires all data points to be known in advance. Put differently, when data arrives iteratively the model must be recomputed at every step and complexity increases steadily with the number k of data points. Fortunately, there are iterative least squares versions for linear approximation.

2.3.2 Online Approximation with Recursive Least Squares

The iterative version of OLS is the so called Recursive Least Squares (RLS) method [3, 41], where the underlying model is updated for every sample individually, but without explicitly storing all data points. This way estimates are already available for few data points and the model continuously improves with incorporation of more samples.

Again, a linear model is considered for now, though OLS and RLS can be applied to other models as well. Rewriting Equation (2.8) into vector notation yields an n-dimensional linear model

$$h(\boldsymbol{x}) = \boldsymbol{\beta}^T \boldsymbol{x}, \tag{2.9}$$

where $x_{n+1} = 1$ and β_{n+1} is the intercept. The approximation goal is to minimize the sum of squared errors over all samples $(\boldsymbol{x}, y)_{1 \leq j \leq k}$, that is, minimization of

$$\sum_{j=1}^{k} \lambda^{k-j} \varepsilon_j^2, \tag{2.10}$$

where $\varepsilon = y - h(\boldsymbol{x})$ is the model error for a single sample and $0 \ll \lambda \leq 1$ is a forgetting factor that assigns exponentially less weight to older samples. The parameter λ can be ignored for now, that is, set to one.

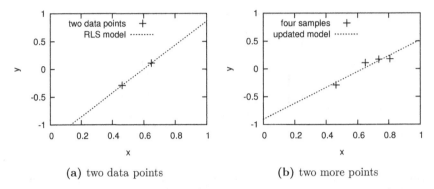

(a) two data points   (b) two more points

Figure 2.3: The RLS algorithm iteratively includes the samples. (a) Two data points can be perfectly fitted with a straight line. (b) Adding two more points reveals that the underlying function is non-linear. The linear RLS model incorporates the data piece by piece.

Without prior knowledge, gradient and intercept are initialized to zero. When a new sample arrives, the gradient must be corrected to account for the new data. This can be seen as a *recursive* process, based on the model at iteration $k - 1$. Generally, model correction can be written as $\beta_{\mathrm{new}} = \beta_{\mathrm{old}} + \varepsilon g$, where $\varepsilon = y - h(x)$ is the a priori model error for the current sample and the so called *gain* g specifies how the model is adapted. The key is to compute a suitable gain such that the model converges to the optimal gradient as exemplified in Figure 2.3. The next paragraph roughly sketches the idea in RLS before details of the math are given.

In order to determine a suitable gain factor g, RLS stores an estimate of the *inverse covariance matrix* P that essentially defines the relevance of inputs for model updates. Initially, any sample should have a strong impact on the model correction and therefore this matrix is initialized as σI, where I is the identity matrix and a large σ defines the update rate during the first iterations. When many samples have been seen in some direction, the matrix P is shrunk in that direction. Thus, overrepresented regions produces small model corrections (low information gain) while rarely sampled regions have a stronger impact (high gain) on the model update. Since the gain is multiplied with the error, the other factor is prediction accuracy. When the model is accurate anyway, no model correction is required. Large errors, on the other hand, result in strong updates.

Detailed derivation of the presented formulas can be found in the literature [3, 41]. A brief summary of the math can be given in three steps. Let (x, y) be the current sample. If an intercept is desired, the input is augmented by a constant entry $x_{n+1} = 1$. First, the gain factor $\boldsymbol{g}$ is computed as

$$\boldsymbol{g} = \frac{1}{\lambda + \boldsymbol{x}^T \boldsymbol{P} \boldsymbol{x}} \boldsymbol{P} \boldsymbol{x} \,, \tag{2.11}$$

where $\boldsymbol{x}^T \boldsymbol{P} \boldsymbol{x}$ defines the overall relevance of the current input $\boldsymbol{x}$ while $\boldsymbol{P} \boldsymbol{x}$ defines the relevance for individual weights β_i, that is, the relevance for each dimension i of the linear model. The computed gain vector $\boldsymbol{g}$ is used in conjunction with the a priori error $\varepsilon = y - h(\boldsymbol{x}) = y - \boldsymbol{\beta}^T \boldsymbol{x}$ to update the linear weights

$$\boldsymbol{\beta} = \boldsymbol{\beta}_{\text{old}} + \varepsilon \, \boldsymbol{g} \,. \tag{2.12}$$

Updating the inverse covariance matrix

$$\boldsymbol{P} = \frac{1}{\lambda} \left(\boldsymbol{P}_{\text{old}} - \boldsymbol{g} \boldsymbol{x}^T \boldsymbol{P}_{\text{old}} \right) \tag{2.13}$$

is the last step which incorporates knowledge about the distribution of inputs seen so far.

The RLS algorithm expects zero mean inputs, that is, $\sum x_j = 0$, as the covariance matrix is a zero centered geometric representation. Furthermore, the noise on function values – if any – is assumed to be zero mean. Finally, the function is assumed to be *static*, that is, the underlying function is not changing dynamically over time. If the function is *dynamic*, a forget rate $\lambda < 1$ helps to continuously adapt to a varying environment. When specific assumptions about the noise can be made or a forward model of the function dynamics are known, an advanced approximation method such as Kalman filtering [51, 3, 41], which is an extension of RLS, should be applied instead.

2.4 Non-Parametric Regression

Instead of fitting data to a predefined function, arbitrary models can be built from the data in various ways. While learning takes considerably longer, the function does not need to be specified in advance. Therefore *interpolation* approaches, in particular Gaussian Processes for regression are described. Artificial Neural Networks are another member of non-parametric regression methods and inherently provide a plausible way to describe brain functionality.

2.4.1 Interpolation and Extrapolation

A rather simple way to approximate a function surface from a *finite* set of samples is interpolation, where predictions are computed as a function of adjacent

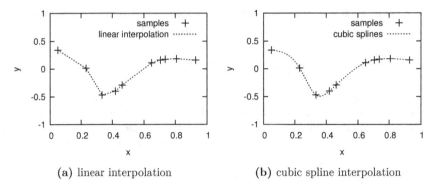

(a) linear interpolation   (b) cubic spline interpolation

Figure 2.4: Interpolation is an intuitive and simple way to produce a function surface within available samples. (a) Linear interpolation connects each sample with a straight line and produces a jagged surface. (b) Smooth interpolation via cubic splines yields a two-times differentiable surface.

data points. For example, if adjacent samples are connected with straight lines (see Figure 2.4a), the model is not only simple but also provides a good intuition of the underlying function. However, the resulting surface is not smooth.

When samples are instead connected with low order polynomials, the surface becomes smooth as illustrated in Figure 2.4b. Here the so called *cubic splines* [6], that is, third order polynomials, are applied. For $k + 1$ samples, the $4k$ unknowns of the k cubic polynomials are specified by the following conditions: First, each polynomial h_i must pass through the two adjacent data points y_i and y_{i+1}, that is,

$$h_i(x_i) = y_i \tag{2.14}$$

$$h_i(x_{i+1}) = y_{i+1}, \tag{2.15}$$

which yields $2k$ equations. Next, the first and second derivatives must match for interior points,

$$h_i'(x_{i+1}) = h_{i+1}'(x_{i+1}) \tag{2.16}$$

$$h_i''(x_{i+1}) = h_{i+1}''(x_{i+1}), \tag{2.17}$$

which yields another $2(k - 1)$ equations, that is, a total of $4k - 2$ equations. Two more conditions are required and typically it is required that the second derivatives are zero at the endpoints. Those equations can be solved for the four polynomial parameters resulting in a symmetric, tridiagnonal system of equations that yields a unique solution for the cubic splines.

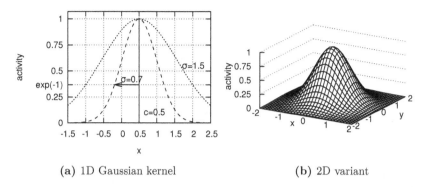

(a) 1D Gaussian kernel   (b) 2D variant

Figure 2.5: Illustration of Gaussian kernels used as a distance metric. (a) The activity of two kernels with $\sigma = 0.7$ and $\sigma = 1.5$ exponentially decreases from the center c, while the maximum activity of one is excited at the center. The width σ defines the distance from center to the inflection point of the activity function, which occurs at a height of $\exp(-1)$. (b) A two-dimensional kernel with $\sigma = 1$ centered on the origin.

2.4.2 Gaussian Process Regression

Another very powerful approach is Gaussian Process Regression (GPR) [67], where not only adjacent data points contribute to the estimated surface, but instead all known samples contribute to the prediction. The rough idea is that samples close to a target should have a stronger impact (of course), but only in the light of all other samples the true trend can be seen.

Therefore, GPR uses a distance metric (also known as kernel function or covariance function) that specifies the relation between samples. Commonly, *Gaussian* kernels are used, that is, an exponential distance metric

$$\phi(\boldsymbol{x}, \boldsymbol{c}) = \exp\left(\frac{-\|\boldsymbol{x} - \boldsymbol{c}\|^2}{\sigma^2}\right) \tag{2.18}$$

from input $\boldsymbol{x}$ to another point $\boldsymbol{c}$, where the quadratic distance is scaled by the width σ of the kernel and the exponential of the negative term yields a Gaussian bump as illustrated in Figure 2.5. Thus, samples far away from a point of interest $\boldsymbol{c}$ have a low influence, while closer samples strongly affect the outcome.

Given k samples $(\boldsymbol{x}_j, y_j)$, the prediction at query point $\boldsymbol{x}_*$ can be formulated as a linear combination of kernels as

$$h(\boldsymbol{x}_*) = \sum_{j=1}^{k} \alpha_j \phi(\boldsymbol{x}_j, \boldsymbol{x}_*) \,. \tag{2.19}$$

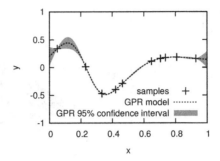

Figure 2.6: GPR provides a good model, even with few samples. Furthermore, the confidence interval clearly indicates points of uncertainty.

The weights α_j are computed in an OLS fashion as

$$\boldsymbol{\alpha} = (\Sigma + \gamma^2 I)^{-1} \boldsymbol{y}, \qquad (2.20)$$

where the $k \times k$ covariance matrix has entries $\Sigma_{ij} = \phi(\boldsymbol{x}_i, \boldsymbol{x}_j)$ and γ^2 is the expected independent and identically distributed noise.

The algebraic path sketched above fits well into this thesis, but is rarely found in the GPR literature. Instead, a probabilistic view is preferred: In addition to the predicted (mean) value a variance is computed. The prediction then becomes a normal distribution with mean

$$\boldsymbol{q}^T \left[\Sigma + \gamma^2 I\right]^{-1} \boldsymbol{y}, \qquad (2.21)$$

where $q_j = \phi(\boldsymbol{x}_j, \boldsymbol{x}_*)$ is the covariance between the j-th sample and the query point. The variance of the predictive distribution is given as

$$\phi(\boldsymbol{x}_*, \boldsymbol{x}_*) - \boldsymbol{q}^T \left[\Sigma + \gamma^2 I\right]^{-1} \boldsymbol{q}. \qquad (2.22)$$

The resulting GPR model for the data set used so far is visualized in Figure 2.6. GPR allows for sophisticated statistical inference – at the cost of an inversion of the $k \times k$ covariance matrix, which grows quadratically in the number of samples. A more detailed introduction into GPR goes beyond the scope of this thesis and the interested reader is referred to [67].

2.4.3 Artificial Neural Networks

Inspired by the capabilities of the human brain, interest in simulated neurons dates back to 1943 [61]. A single neuron represents a kind of function, as it receives inputs and produces some kind of output. Thus, a network of neurons, so called Artificial Neural Network (ANN), is clearly related to FA as well. Furthermore, biological plausibility comes for free with such models, when necessary simplifications to simulate neural activity in a computer are ignored.

The Cybenko theorem [23] states that any continuous function can be approximated with arbitrary precision on a (compact) subset of $\mathbb{R}^n$ with a particularly simple type of ANN. On first sight, this theorem might make ANN the first choice for the present thesis. However, arbitrary precision may come at the cost of arbitrary computational resources and the theorem is not constructive in that it does not provide the optimal number of neurons for a given task. Furthermore, the mere existence of an *optimal* network does not imply that a learning algorithm converges to that solution on a complex problem.

There is a vast number of different network types including, but not limited to, simple feed forward networks [61], recurrent neural networks [36], and spiking neural networks [32]. The most attractive type for FA in the present work, tough, are Radial Basis Function Networks (RBFNs) [40]: They fall into the Cybenko category of ANNs and arbitrary approximation precision is possible. However, in contrast to other networks, such as multilayer perceptrons, the *globally optimal* solution for a given data set can be computed in a simple fashion.

Radial Basis Function Networks

RBFNs [40] belong to the family of so called feed forward neural networks that do not have cyclic connections. They can be described as a three layer architecture composed of an input layer, a hidden layer, and an output layer. Inputs are propagated to the hidden layer, where a non-linear activation function determines the activity of the hidden neurons. Every hidden neuron has a simple model and votes according to its activity and model. The votes are aggregated and provide the estimated function value for the given input as depicted in Figure 2.7.

Formally, the prediction with m hidden neurons is a weighted sum

$$h(\boldsymbol{x}) = \frac{\sum_{r=1}^{m} \phi_r(\boldsymbol{x})\lambda_r(\boldsymbol{x})}{\sum_{r=1}^{m} \phi_r(\boldsymbol{x})}, \qquad (2.23)$$

where ϕ_r represents the r'th radial basis function and λ_r are local models, be it a plain weight or a linear model. Here, the normalized version is considered, that is, the weighted vote from all models is divided by the sum of all weights.

Typically a Gaussian kernel (cf. Equation (2.18) and Figure 2.5) is used as the radial basis function

$$\phi_r(\boldsymbol{x}) = \exp\left(\frac{-\|\boldsymbol{x} - \boldsymbol{c}_r\|^2}{\sigma^2}\right), \qquad (2.24)$$

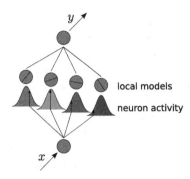

Figure 2.7: Each neuron of an RBFN receives the input x and excites a certain activity. The contribution of a hidden neuron to the output is defined by its local model which is, in the simplest case, just a single weight. The final output is a activity-weighted sum of the individual models.

where the width of all kernels in a RBFN is usually a fixed, global value σ, while the centers c_r are distributed according to the problem at hand. Non-linearity of the activation function is an important property here, as the *linear* combination of *linear* models collapses to a single linear model.

Learning, which takes place at the level of local models λ_r, often referred to as *weights*. The simplest RBFN approach uses a single weight $\lambda_r(x) = \alpha_r$ independent of the actual input and Equation (2.23) becomes

$$h(x) = \frac{\sum_{r=1}^{m} \phi_r(x)\,\alpha_r}{\sum_{r=1}^{m} \phi_r(x)}\,, \tag{2.25}$$

Those weights can then be optimized with respect to the squared error. In contrast to other ANNs, where the popular *back-propagation* algorithm computes suitable weights, the *globally optimal* weights for an RBFN can be found by solving a linear system of equations. Thus, this is closely related to OLS, where linear weights were computed for a linear model. Here, linear weights are computed as well – but the non-linear activation function allows for non-linearity in the final RBFN model.

Given a data set of k samples (x, y) with $x \in \mathbb{R}^n$, the model for all samples can be written in matrix notation as

$$\boldsymbol{H\alpha} = \begin{pmatrix} H_{11} & \cdots & H_{1m} \\ \vdots & \ddots & \vdots \\ H_{k1} & \cdots & H_{km} \end{pmatrix} \begin{pmatrix} \alpha_1 \\ \vdots \\ \alpha_m \end{pmatrix} = \begin{pmatrix} y_1 \\ \vdots \\ y_k \end{pmatrix} = \boldsymbol{y}\,, \tag{2.26}$$

where the matrix entry at index j, q is the normalized activity of the q'th kernel on the j'th sample

$$H_{jq} = \frac{\phi_q(\boldsymbol{x}_j)}{\sum_{r=1}^{m} \phi_r(\boldsymbol{x}_j)} \tag{2.27}$$

The former Equation (2.26) is equivalent to an ideal, that is, zero-error model evaluated at all inputs. Thus, the optimal weights $\boldsymbol{\alpha}$ can be derived by rewriting the above equation into $\boldsymbol{\alpha} = \boldsymbol{H}^\dagger \boldsymbol{y}$, where $\boldsymbol{H}^\dagger = (\boldsymbol{X}^T \boldsymbol{X})^{-1} \boldsymbol{X}^T$ denotes the Pseudoinverse matrix. This is the same strategy as used for the OLS solution in Equation (2.7) or for GPR in Equation (2.20), and therefore another example for using OLS with a non-linear model.

Importantly, computing the weights this way does not necessarily result in a zero-error model, but instead minimizes the MSE for the given RBFN. Placing one kernel on top of each sample is a simplified from of GPR. However, analogously to OLS and GPR this approach requires all samples in advance and is not suited for online approximation on infinite data streams.

Iterative Update Rules for RBFNs

An alternative to the batch-training is a gradient descent that continuously adapts the model with a certain learning rate. Given sample $(\boldsymbol{x}, y)$, a simple update rule for one neuron is

$$\alpha = \alpha_{\text{old}} + \delta \underbrace{\phi(\boldsymbol{x})}_{\text{activity}} \underbrace{(y - h(\boldsymbol{x}))}_{\text{error}}, \tag{2.28}$$

where $0 < \delta < 1$ defines the learning rate. Training an RBFN with six kernels and a learning rate of $\delta = 0.5$ over 15000 iterations on the given data set (see Figure 2.8) yields satisfactory results.

While the OLS solution depicted above optimizes all weights globally, the gradient descent method in Equation (2.28) trains each neuron locally – not taking information from neighboring neurons into account. Put differently, here each neuron minimizes its own error while the OLS method minimizes the sum of (squared) errors from *all* neurons. The global optimization naturally produces a more accurate model, while gradient descent is applicable to infinite data streams.

Up to now, nothing has been said about the shape of the radial basis functions, that is centers $\boldsymbol{c}_r$ and width σ, and this is the crux of RBFN. On a finite data set one kernel can be set on top of every data point, which is a basic form of interpolation with GPR. Alternatively a reduced number of centers can be chosen randomly among inputs, distributed uniformly, or even completely random in the input space.

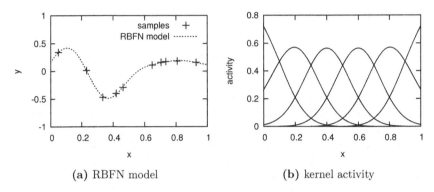

(a) RBFN model   (b) kernel activity

Figure 2.8: A simple six-neuron RBFN is trained for 15000 iterations. (a) The function structure is well approximated. (b) Six Gaussian kernels are uniformly distributed with $\sigma = 0.2$. Normalized activity is shown here.

While this might give acceptable results, the intuition is that centers and corresponding widths should be tuned as well. In this case, the optimization becomes a two-step procedure: a) choosing centers and widths, b) training. The so called *meta* learning is repeated until performance is satisfactory. The following section introduces two algorithms that interweave the optimization of shape, that is, clustering and training of the local models.

2.5 Local Learning Algorithms

RBFNs fit quite well to the requirements for this thesis stated in Section 1.4. It is based on a population of neurons that cooperatively model a non-linear function with several local, simple sub-models. Two crucial factors affect the performance of RBFN *jointly*: the shape of the radial basis functions and the weights of the local models. The latter can be trained in a rather short time scale. However, to assess (or optimize) the shape of the neuron activation functions, the weights have to be trained as well – therefore optimization of the shape requires considerably more effort.

Algorithms that train several simple, parametric models in a certain locality *and* learn the locality – that is, a clustering of the input space – are henceforth called *local learning* algorithms. Where RBFNs adapts its weights based on the global prediction error (average from all neurons), a local learner updates it's parametric models on their individual errors. The most demanding requirement for this class of algorithms is *online* functionality, that is, function approximation on a continuous stream of samples. For finite data sets, closed form solutions such as GPR can provide high accuracy. Iterative solutions, on the other hand, can hardly reach the accuracy of a closed form solution and, additionally, require exhaustive training. In the following, two such algorithms are briefly introduced.

2.5.1 Locally Weighted Projection Regression

Similar to RBFN the so called Locally Weighted Projection Regression (LWPR) [75, 91] approximates a non-linear function surface by several locally linear models. LWPR also applies Gaussian kernels to cluster the input space. However, the shape of the kernels is optimized concurrently during training of the linear models. Since this is done iteratively, on a stream of samples, the approach becomes rather complex compared to RBFNs. The ideas are outlined briefly.

First, the local models applied in LWPR are linear, but neither trained by RLS nor by a simple gradient descent as done in Equation (2.28). Instead, the so called Partial Least Squares (PLS) method is applied, which does not regress the full n-dimensional input against the output, but instead projects the full input space onto the $l < n$ most relevant dimensions. This reduced input space is then used to estimate a linear relation to the output. While it allows to treat high-dimensional, partially irrelevant data, PLS also requires another parameter: the desired dimension l of the reduced input space. In LWPR this parameter is incremented until the performance gain can be neglected (which turns out to be another parameter).

The major difference to RBFNs, though, is the on-demand creation of kernels and the optimization of kernel *shape* and *size* during training. When a new sample is fed into LWPR and no kernel elicits an activation greater than some threshold, a new kernel centered on the new input is created. The shape of kernels is then optimized due to a cost function close to the least squares sense.

However, when the shape is optimized to minimize the squared error, then the obvious solution are very small kernels, so small that one kernel is responsible for one data point. With an increasing amount of data, an ever increasing number of very small kernels is required and consequently a slightly different approach is taken. Let ϕ be a single kernel with its local model h to approximate samples $(\boldsymbol{x}, y)_i$. Instead of minimizing the sum of *all* squared errors

$$\frac{\sum \phi(\boldsymbol{x}_i)(y_i - h(\boldsymbol{x}_i))^2}{\sum \phi(\boldsymbol{x}_i)}, \tag{2.29}$$

the leave one out cross validation [75]

$$\frac{\sum \phi(\boldsymbol{x}_i)(y_i - h(\boldsymbol{x}_{i,-i}))^2}{\sum \phi(\boldsymbol{x}_i)} \tag{2.30}$$

is used, where the index $i, -i$ indicates that the i-th sample is trained on all other data points $j \neq i$. Still, this yields infinitely small kernels and therefore a shrinkage penalty is added

$$\frac{\sum \phi(\boldsymbol{x}_i)(y_i - h(\boldsymbol{x}_{i,-i}))^2}{\sum \phi(\boldsymbol{x}_i)} + \gamma \sum D_{jk}, \qquad (2.31)$$

where D_{jk} is one entry of the kernel matrix and a smaller entry refers to a greater radius. Minimizing the latter term yields infinitely large kernels, minimizing the former results in infinitely small kernels, and the parameter γ defines the balance. Minimizing both means reduction of the squared error with reasonably sized kernels. Details of the incremental update rules that approximate this optimization can be found in [75, 91].

2.5.2 XCSF – a Learning Classifier System

An alternative route is taken by the XCSF [98, 99] algorithm. The naming conventions are – again – different but the idea remains the same. For now, only a brief introduction is given, as the algorithm is explained in depth later.

XCSF evolves a population of rules, where each rule specifies a kernel accompanied by a local model. The rules jointly approximate a function surface. Moreover, it works online, that is, iteratively. Local models are typically linear and trained via RLS; the kernels can be described by arbitrary distance metrics.

In contrast to LWPR however, the shape and size of kernels is not adapted by a stochastic gradient descent, but instead by a Genetic Algorithm (GA). Therefore, the cost function is termed fitness function and its explicit derivative is not required, as a GA does not directly climb the gradient but instead samples some individuals around the current point and picks its path based on the evaluated surrounding. This allows for more complicated cost (fitness) functions, where explicit derivatives are not easily available.

Instead of introducing a penalty term, the error is not minimized to zero but to a certain target error. This gives a clear idea of the desired accuracy and resolves the balancing issue of parameter γ in Equation (2.31) for LWPR. However, other tricks are required to achieve a suitable population of kernels, including a *crowding* factor that prevents that the majority of kernels hog to simple regions of the function surface but instead spread over the full input space. A detailed algorithmic description is given later.

2.6 Discussion: Applicability and Plausibility

The different approaches to FA introduced in the previous sections are now inspected for their *applicability* to the problem statement of this thesis as well as for the *biological plausibility*. Lets recall the requirements for FA stated in Section 1.4:

- online learning,

- with feasible computational complexity,

- using some neuron-like structures in a population code,

- without putting too much prior information into the system.

When the underlying function type is known (e.g. linear, polynomial, exponential, etc.), only the proper parameters of that function (e.g. $ax^2 + bx + c$) have to be found. The so called *curve fitting*, *model fitting*, or *parametric regression* methods find those parameters a, b, c as outlined in Section 2.3. Iterative variants with acceptable computational complexity are available as well, e.g. RLS. Furthermore, a neural implementation of RLS is possible [95]. However, when the function type is hard-coded into the learner – for example, that would be a combination of multiple sine waves for a kinematic chain – the biological plausibility is questionable. Humans can not only work with kinematic chains they are born with, e.g. arms and legs, but also learn to use new, complex tools that have effects quite different from sine waves: for example, driving a car.

An alternative, rather simple approach to FA without particular assumptions on the function type is *interpolation*. Even simple spline-based models can result in suitable models of highly non-linear function surfaces. GPR is the statistically sound variant that shows high accuracy even for small sample size at the cost of a quadratic computational complexity in the number of samples. However, all such approaches require to store the full data set and therefore apply to finite data sets only. Even if only a subset of a continuous data stream (that a brain receives every day) is stored, the biological plausibility is questionable: Are kinematic models stored in an neural look-up table that is queried for movement execution? It is reasonable to assume a higher level of storage abstraction for population codes, where each neuron is responsible for a certain surrounding not just a single point in space.

As ANNs are inspired by the brain structure, they are inherently plausible in the biological sense. In particular, the RBFN is a suitable candidate for the present work: A population of neurons jointly models a non-linear function surface. Each neuron responds to a certain area of the input space defined by a kernel function. In its area of responsibility a simple local model is learned by an iterative gradient descent. Thus the method works online and, furthermore, with a low computational demand. The major drawback of RBFNs is the missing training possibilities for the kernels.

This issue is addressed by two *local learning* algorithms, namely LWPR and XCSF. Similar to RBFNs those algorithms iteratively model non-linear functions with several local models whose area of responsibility is defined by a kernel. This can be seen as a population code, a population of neurons that learns a particular mapping from one space onto another. The computational demand highly depends on the number of neurons or kernels, which does not matter in a parallelized environment such as the brain, where neurons fire asynchronously, independently of each other.

There are many other FA techniques available, and some of them may be extended or modified to fit the desired framework outlined here. However, the simple RBFN and its more sophisticated brothers LWPR and XCSF naturally fit the needs for a biologically plausible learning framework. The next chapter analyzes the common features of those algorithms in depth to not only foster a better understanding but eventually to lay the foundation for new Machine Learning (ML) algorithms that are well-suited to explain cognitive functionality.

3 Elementary Features of Local Learning Algorithms

Locally Weighted Projection Regression (LWPR) and the learning classifier system XCSF share a common *divide and conquer* approach to approximate non-linear function surfaces by means of three elementary features:

Clustering A complex non-linear problem is broken down into several smaller problems via kernels. The kernel structures are further optimized for accurate approximations.

Local Models A simple, parametric regression model is trained in the locality of each cluster, e.g. locally linear models.

Inference Predictions are computed as a *weighted* sum of the local models, where models close to the corresponding input have more influence than those far away.

This approach is termed *local learning* here, where not only the local models learn but the locality itself may be learned as well, that is, size and shape of the kernels. This stands in contrast to so called *kernel regression* methods that focus on computing linear weights in the feature space of the kernel without explicitly optimizing the structure of kernels itself (e.g. Radial Basis Function Network (RBFN)).

The chapter at hand first takes a detailed look at the three features individually. Different kernel structures – from simple to complex – are discussed and their properties outlined in Section 3.1. In turn, Section 3.2 briefly recapitulates constant and linear models. Third, *weighting* of the local models is explained in Section 3.3. Finally, the *interaction* between kernel structure, local models, and weighting is highlighted in Section 3.4.

Altogether this chapter only describes the model *representation*, but does not yet tackle how the locality, that is, the clustering can be *learned*, which depends on the actual algorithm. The basic RBFN approach does not modify the shape and size, but instead requires fixed parameters. LWPR tries to optimize the kernels by a estimated gradient descent instead. XCSF on the other hand relies on a Genetic Algorithm (GA) to find an optimal clustering without explicitly computing a gradient. While the methods have been sketched in the previous chapter, the full optimization procedure of XCSF, including optimization of the clustering itself, is detailed in the subsequent chapter. For now, the grounding features for such an optimization are discussed.

3.1 Clustering via Kernels

Kernel functions are widely used but often termed differently. For example, Gaussian Process Regression (GPR) [67], support vector machines [12], and kernel density estimation [72] apply kernels. In RBFN [40] the same method is called *radial basis function*, LWPR [92] uses the term *receptive field*, and a so called *classifier* in XCSF [99] can be seen as a kernel as well.

Unfortunately, there is no general consensus on the definition of a kernel. For the present thesis, a kernel

$$\phi : V \times V \to \mathbb{R} \qquad (3.1)$$

represents a *measure of similarity* between two points of the vector space V, often $V \subset \mathbb{R}^n$. The intuition is that points $x_1, x_2 \in V$ close to each other should produce a larger activation than points far away from each other. Put differently, points close to each other in input space are likely to have *similar* function values and, thus, can be approximated by a simple, local model. Apart from this common concept of similarity, the mathematical notation and stated requirements are often different. This section aims to give a unified view on the topic. First, some different definitions for a kernel are presented.

The weakest formulation defines a kernel as an arbitrary non-linear function [40]

$$\phi(\|x_1 - x_2\|) \qquad (3.2)$$

on some norm $\|\cdot\|$, e.g. the Euclidean distance $\|x\| = \sqrt{\sum_{i=1}^n x_i^2}$. One example is the Gaussian kernel from Equation (2.18). Contrary to intuition, the measure of similarity can be reversed as well when ϕ grows with increasing distance of two points. For example, the *multi-quadratic* kernel [40] can be defined as

$$\phi(x_1, x_2) = \sqrt{1 + \frac{\|x_1 - x_2\|^2}{\sigma^2}}, \qquad (3.3)$$

where σ specifies the width of that kernel (c.f. Figure 3.1). Values far apart produce a large activation, which is more a *distance* measure than a *similarity* measure. Furthermore, negative values are possible as well, as for example with the *thin plate spline* kernel [26]

$$\phi(x_1, x_2) = \frac{\|x_1 - x_2\|^2}{\sigma^2} \ln \left(\frac{\|x_1 - x_2\|}{\sigma} \right). \qquad (3.4)$$

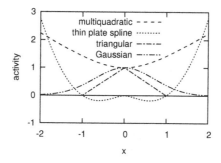

Figure 3.1: Often few restrictions are made for a kernel, only symmetry is implied by the use of a norm. Here, $\phi(x, 0)$ is depicted to illustrate activation functions with respect to zero. In contrast to multi-quadratic and thin plate spline kernels, the Gaussian kernel and the 1-norm measure *locality*, that is, points closer to zero produce a higher activation.

Finally, the norm does not need to be smooth and the triangular kernel [73]

$$\phi(\boldsymbol{x}_1, \boldsymbol{x}_2) = \max\{1 - \frac{\|\boldsymbol{x}_1 - \boldsymbol{x}_2\|}{\sigma}, 0\} \tag{3.5}$$

results in an activation function that is not differentiable. The most common type of kernel, however, is the Gaussian one defined in Equation (2.18), which is displayed in Figure 3.1 together with the above mentioned activation functions.

A more strict specification is used for kernel density estimation methods [72], where the integral of the kernel over the full range

$$\int_{-\infty}^{+\infty} \phi(\boldsymbol{x}) d\boldsymbol{x} \tag{3.6}$$

is required to be positive, finite and, furthermore, the function ϕ is required to be non-negative. Such a kernel inherently specifies a measure of *locality*. Since values close to each other will likely have similar function values (on smooth functions), this definition is more useful than the general definition seen before. In contrast to multi-quadratic, thin plate spline, and triangular kernels, the smooth Gaussian activation function satisfies this condition. For alternatives to this widely used activation function, the interested reader is referred to the fourth chapter of Rasmussen's book on GPR [67].

By altering the norm and its corresponding width parameter σ itself, the geometric shape can be further tuned. The following sections illustrate several geometric shapes based on Gaussian activity and the required parameters. As a kernel is always located at center c to measure the distance to some point x, the notation $\phi(x)$ is short for $\phi(x, c)$.

3.1.1 Spherical and Ellipsoidal Kernels

A spherical kernel centered at $c \in \mathbb{R}^n$ is the simplest formulation with a single parameter: the radius σ, sometimes termed length scale.

$$\phi(x) = \exp\left(-\frac{\|x - c\|^2}{\sigma^2}\right)$$

$$= \exp\left(-\sum_{i=1}^{n} \frac{(x_i - c_i)^2}{\sigma^2}\right) \tag{3.7}$$

If the squared distance from input x to center c equals the squared radius σ, the fraction equates to one, which is the inflection point of the exponential activation function at $\phi(x) = \exp(-1)$.

Ellipsoids

The extension to *ellipsoids* is straight forward: Instead of using the same radius for all dimensions, an individual radius σ_i is applied to the i'th dimension. We rewrite the equation into a convenient matrix notation

$$\phi(x) = \exp\left(-\sum_{i=1}^{n} \frac{(x_i - c_i)^2}{\sigma_i^2}\right)$$

$$= \exp\left(-\sum_{i=1}^{n} (x_i - c_i)\frac{1}{\sigma_i^2}(x_i - c_i)\right)$$

$$= \exp\left(-(x - c)^T \Sigma (x - c)\right), \tag{3.8}$$

where Σ is an $n \times n$ matrix and the diagonal matrix entries are set to $\Sigma_{ii} = 1/\sigma_i^2$ for $1 \leq i \leq n$. The term $(x - c)^T \Sigma (x - c)$ yields the quadratic distance, scaled by the individual radii in each dimension. In n dimensions, such a representation requires n radii to be specified. As an example, a two-dimensional kernel centered at $(0, 0)$ with $\sigma_1 = 0.6$, $\sigma_2 = 0.3$ is depicted in Figure 3.2a.

General Ellipsoids with Rotation

When problem dimensions are not independent, their relevant axes might lie oblique in the input space and such structures cannot be well captured with

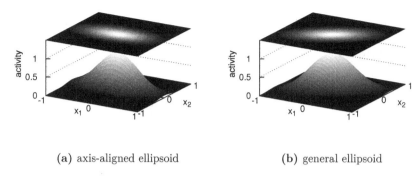

(a) axis-aligned ellipsoid (b) general ellipsoid

Figure 3.2: Illustration of ellipsoidal kernel functions. (a) The radius in x_1 direction is $\sigma_1 = 0.6$ and $\sigma_2 = 0.3$ implies half the width along the x_2 axis. (b) The radii are the same, but the kernel is rotated about its center.

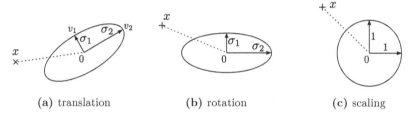

(a) translation (b) rotation (c) scaling

Figure 3.3: The ellipsoidal distance computation consists of three steps. (a) The input is translated to the center of the ellipsoid. Thus, the origin of the coordinate system is located at the center of the ellipsoid. (b) The inverse rotation creates an axis-aligned representation. (c) Inverse scaling yields the unit-sphere, where distance computation is simple.

axis-aligned representations. Therefore, an $n \times n$ rotation matrix V can be included in the distance computation

$$\phi(x) = \exp\left(-(x - c)^T V \Sigma V^T (x - c)\right), \tag{3.9}$$

which then allows for rotated ellipsoids as illustrated in Figure 3.2b. The matrix $V \Sigma V^T$ is an $n \times n$, symmetric, positive semi-definite matrix and consequently also represents a valid covariance matrix.

Altogether the equation combines three elements of an inverse, affine transformation as depicted in Figure 3.3. First, the point x is translated to the center c of the kernel. Second, the inverse rotation $V^T = V^{-1}$ is applied, which puts the point into the *axis-aligned* representation of the original ellipsoid. Third, the inverse scaling is applied to the point and thus, the ellipsoid has become a unit-sphere. Here, a simple dot product yields the quadratic distance. To further illustrate this process, the squared distance computation within Equation (3.9) can be rewritten in

$$(x - c)^T V \Sigma V^T (x - c) = \left((x - c)^T V \sqrt{\Sigma} \right) \left(\sqrt{\Sigma} V^T (x - c) \right)$$
$$= \left(\sqrt{\Sigma} V^T (x - c) \right)^T \left(\sqrt{\Sigma} V^T (x - c) \right), \qquad (3.10)$$

where $\sqrt{\Sigma}$ is a diagonal matrix composed of the inverse radii $1/\sigma_i$ and thus $\Sigma = \sqrt{\Sigma}\sqrt{\Sigma}$. The resulting squared distance is then used with the exponential function to yield a Gaussian shape.

Unfortunately, the complexity of rotation is generally $\mathcal{O}(n^2)$ because an n-dimensional rotation has $n(n-1)/2$ degrees of freedom. Additionally, n radii make up for a total of $n(n+1)/2$ parameters for such a representation. It is non-trivial to find the optimal rotation parameters for a regression problem at hand, but there are much richer options to cluster the input space. Methods to optimize such a complex shape will be introduced in a later chapter.

3.1.2 Alternative Shapes

The use of a quadratic distance divided by squared radii results in an ellipsoid equation. When instead of a quadratic distance from center c the maximum distance of all individual dimensions is used, the contour of the kernel becomes rectangular as shown in Figure 3.4.

While axis-aligned rectangles are useful for discrete input spaces, their application in continuous scenarios disagrees with the idea of similarity that smoothly changes over the input space: A smooth function rarely features an edgy contour.

While geometric shapes are rather intuitive, even with rotation, any type of function that provides some kind of activity could be used to cluster the input space – dropping the idea of similarity and instead allowing for arbitrary activations. For example, a neural network might activate above a given threshold in any fancy shape, hard to describe, but possibly very powerful depending on the problem at hand [47, 59]. Alternatively, a mathematical description of the activity function could be evolved with a Genetic Programming approach [100]. The current work will not deal with those, but stick with axis aligned and general ellipsoids via Gaussian kernels that can be well interpreted and have powerful clustering capabilities.

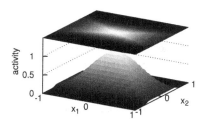

Figure 3.4: A rectangular kernel is obtained by using a max-norm instead of a quadratic one.

3.2 Local Models

The aforementioned kernels define a niche in the input space. The task of fitting a model to samples in that niche is the topic of this section. An iterative approach – as opposed to batch algorithms – is required because samples are streamed piece by piece.

First off, the simplest approach is to compute an average of all samples seen

$$h = \frac{1}{k} \sum_{j=1}^{k} y_j , \qquad (3.11)$$

which represents a *constant* model. Here the term "constant" refers to the model prediction with respect to its input space, but does not refer to time, since the model may change in every step. Obviously, this is not the best thing to do – but on the other hand it is fast and completely independent of the problem dimensionality. An iterative update rule for a so called running mean h can be simply written as

$$h_{j+1} = h_j + \frac{1}{j+1} \left(y_{j+1} - h_j \right) , \qquad (3.12)$$

where y_{j+1} is the latest function value to be incorporated into the previous mean h_j. Alternatively, a fixed learning rate $0 < \delta \ll 1$ can be used as

$$h_{j+1} = h_j + \delta \left(y_{j+1} - h_j \right) , \qquad (3.13)$$

which is sometimes called *delta* rule or *Widrow-Hoff* rule. One the one hand, the Widrow-Hoff model cannot converge unless the data is indeed constant.

On the other hand, the forgetting of old data allows adaptation in dynamic environments.

The locality defined by the kernel can be incorporated in different ways. For example, one could minimize the squared error in the area, where the kernel elicits higher activation than any other kernel. This concept does not change the update rule itself, but merely defines *when* an update is triggered. Instead of using a hard border for the niche, the local model could receive *every* sample of the input space, but the update strength may be weighted by activity of its corresponding kernel ϕ.

$$h_{j+1} = h_j + \phi(x_{j+1})\delta\left(y_{j+1} - h_j\right), \tag{3.14}$$

Since the Gaussian kernel activity decays exponentially, this yields a very localized model with slightly superior accuracy at the center compared to the former approach that aims for a more generalized model with a larger area of influence.

While moving averages as described above provide a simple model, the extension to linear models offers an accuracy improvement with little computational overhead. Section 2.3.2 described the Recursive Least Squares (RLS) method that iteratively approximates the given data with, e.g., a linear model. A weighted update is realized by multiplication of the gain factor g with the weight, e.g. activity ϕ. Even more sophisticated models can be plugged in, such as polynomial RLS models [55]. This shifts the approximation power from a fine grained problem-appropriate clustering towards a complex local model.

3.3 Inference as a Weighted Sum

When predictions are requested or the whole function surface is to be plotted, the model is queried at arbitrary points in input space – also called *inference*. With kernels and their local models this basically means *weighted recombination* of the local models.

As with the update rules many roads lead to Rome. A simple approach without any weighting picks the node with highest activity of the m kernels and uses its model h_r as the prediction

$$p(x) = h_r(x) \quad \text{where } r = \arg\max_r \phi_r(x) \tag{3.15}$$

Alternatively the weighting can be an activity weighted sum, that is

$$p(x) = \frac{\sum_{r=1}^m \phi_r(x)h_r(x)}{\sum_{r=1}^m \phi_r(x)} \tag{3.16}$$

with ϕ_r being the r'th kernel and h_r the corresponding local model. The denominator normalizes the weighted prediction. This method is applied in RBFN as well as LWPR.

Furthermore, mixed approaches are possible as well. All kernels that elicit activation greater then some threshold α are selected. In turn, the selected models are weighted by their accuracy, that is, an inverse of the prediction error ε.

$$M = \{r \mid \phi_r(\boldsymbol{x}) \geq \alpha\}$$

$$p(\boldsymbol{x}) = \frac{1}{\sum_{r \in M} \frac{1}{\varepsilon_r}} \sum_{r \in M} \frac{h_r(\boldsymbol{x})}{\varepsilon_r} \qquad (3.17)$$

This might be particularly useful when certain models are well trained but others are not, as is the case in XCSF as detailed later.

3.4 Interaction of Kernel, Local Models, and Weighting Strategies

When kernels divide the input space, local models act on individual clusters, and a weighted sum represents the final prediction, the composite result is not immediately clear. This section sheds light on the *interaction* of those three features with some simple examples.

The Cybenko theorem [23] states that *arbitrary accuracy* is possible for an RBFN. This essentially requires a large number of neurons, that is, pairs of kernel and local model. Suppose a RBFN with m neurons approximates a linear function with constant weights. With $m \to \infty$ the accuracy converges to 100%. Another way to improve accuracy is the use of sophisticated local models. A single neuron equipped with a *linear* model would reach 100% accuracy on a linear surface as well. Last but not least, two neurons with constant models can be placed on the left-most and right-most inputs. A linear weighting between those two neurons would perfectly resemble the linear function in between.

All three features – kernel, local model, and weighting strategy – *jointly* affect the final prediction outcome. A more detailed example illustrates the different effects. Suppose a simple max-weighting is used, where only one kernel at a time contributes to the model, and the simplest model possible: a constant value for the corresponding kernel as shown in Figure 3.5a. Here, the kernels are distributed uniformly.

Increasing the number of neurons would clearly improve the prediction. However, also the location and size of the kernels is important. The function seems more difficult on the left-hand side and a fine-grained clustering on the left improves the accuracy as illustrated in Figure 3.5b. Moreover, Figures 3.5c and 3.5d show that even higher accuracy can be achieved with linear models.

Finally, when a sophisticated weighting strategy is chosen the prediction accuracy can be further improved. Figure 3.6 depicts two Gaussian kernels with

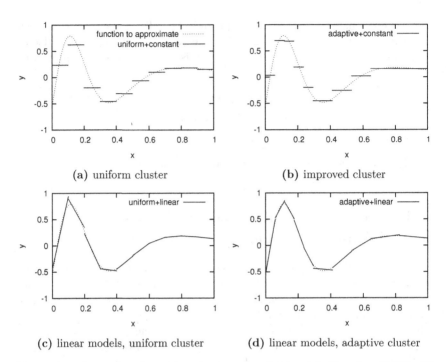

(a) uniform cluster   (b) improved cluster

(c) linear models, uniform cluster   (d) linear models, adaptive cluster

Figure 3.5: Approximation with a simple max-weighting (cf. Equation (3.15)). The underlying function (dashed) is unknown. (a) A uniform cluster with constant models results in large errors on the left-most side. Here the overall Root Mean Squared Error (RMSE) is 0.1432. (b) When more kernels cover the *difficult* left part and fewer resources are assigned to the simple right half, the RMSE reduces to 0.0982. (c) The RMSE drops to 0.0266 when linear models are plugged in and an even lower error of 0.0132 is achieved with both linear models and a suitable clustering in (d).

their local models weighted by kernel activity as described in Equation (3.16). This results in a smooth transition from one local model to the other and reduces the importance of a model that perfectly approximates the underlying data.

To sum up, all three features can improve prediction quality and they should be chosen problem dependent. The widely used Gaussian kernel function can be used to produce spheres, axis-aligned ellipsoids, and even general ellipsoids to cluster an input space. Gaussian kernels are smooth and reflect the notion of *similarity* with a maximum activation at the kernel center, exponentially decaying with increasing distance.

The local model put upon the kernel structure can be simply constant as a moving average, but locally linear models require little more computational

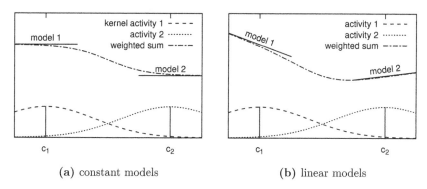

(a) constant models   (b) linear models

Figure 3.6: Weighting of two local models by a Gaussian activity function. (a) Two constant models produce a smooth surface with proper weighting. (b) Linear models further improve the model expressiveness.

resources compared to an improved expressiveness. While higher order models are possible as well, their accuracy benefit might be negligible if the underlying data does not have truly polynomial features. Furthermore, the training time for each local model increases which, in turn, increases the overall training time of a local learning algorithm.

Smooth transitions from one local model to the next can be achieved with a smooth weighting strategy, e.g. based on the Gaussian activity. Alternatively a trust measure such as the average prediction error of a local model can be used to balance multiple local predictions which is exemplified in the following chapter, which explains the XCSF algorithm in detail.

4 Algorithmic Description of XCSF

With the three elementary features at hand, it is now time to take a detailed look at one such local learning algorithm that not only optimizes the weights between kernels, but also optimizes the kernel *shape* to further improve prediction accuracy. The XCSF algorithm is a so called Learning Classifier System (LCS), where a Genetic Algorithm (GA) optimizes a population of rules. A rule consists of a kernel with particular location, shape, and size, and a local model for the corresponding subspace.

Initially, LCS were envisioned as *cognitive* systems that interact with some kind of environment [44, 43]. One of the most successful LCS is the XCS system [96], that was modified for Function Approximation (FA) as well [99], then called XCSF. An exhaustive review about LCS variants was written by Urbanowicz and Moore [89]. Since XCSF can approximate non-linear, multi-dimensional functions online from iterative samples, the system is well suited to our problem of learning robot models for live control. As before with Radial Basis Function Networks (RBFNs) it is a particular combination of clustering, local models, and a weighting strategy. However, as opposed to RBFN the clustering is optimized intermittently with the training of local models.

Since a kernel is always equipped with a local model, the combination of those is herein termed a Receptive Field (RF), which simplifies explanations and also highlights the similarity to neurons. In the LCS literature, a RF is termed *classifier*. XCSF has a population of RFs that are optimized over time to accurately model the underlying data. Each RF defines its shape as discussed in Section 3.1 and a local model approximates the data within its radius of responsibility. Together, many RFs model the whole surface of the function by means of a particular weighting strategy.

4.1 General Workflow

The input for XCSF is a stream of function samples $(x, y)_t$ where $y = f(x)$ at time t and the workflow consists of four elementary steps:

Matching The population is scanned for *active* RFs, sometimes called *match-set*, that consists of all RFs with a certain minimum activity. If no RF is responsible for the current input x, a new RF with its center equal to x and a random shape is created.

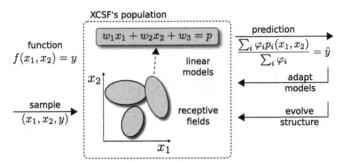

Figure 4.1: Workflow of the XCSF learning classifier system on a 2D function. Three RFs match a sample input (x_1, x_2) and a linear prediction is computed from each model. The weighted combination yields a global prediction that is used to adapt local models and evolve the global structure of the population.

Weighted Prediction Active RFs compute individual predictions $p_i(x)$ using their local models and a global prediction p is formed as a weighted sum of individual predictions.

Local Model Adaptation Given the true function value y a prediction error $\varepsilon_i = \|y - p_i(x)\|$ is computed for each local model. The local, current prediction error ε_i is then used to adapt the models in a least squares fashion. This is a local, micro-scope optimization to incorporate the current data point in the local models knowledge.

Global Structure Evolution Finally, the average prediction error of individual RFs is used to further refine the actual clustering, that is, the structure, shape, and size of the RFs. Basically, the GA tries to minimize the prediction error, but several subtle issues need to be taken care of. This is the macro-scope optimization.

Figure 4.1 illustrates the workflow on a two-dimensional function. The following sections explain the details of each step.

Let $f(x) = y, x \in \mathbb{R}^n, y \in \mathbb{R}^m$ be an n-dimensional function with m-dimensional real-valued outputs. The function is sampled iteratively, that is, at time step t a new input x is handed to XCSF. This could be a random point of the input space, a particular sequence from an underlying data set, or even a trajectory. While there is no restriction for the distribution of inputs, the distribution has an implicit effect on the performance as we will see later. XCSF starts with an initially empty population $P = \emptyset$.

4.2 Matching, Covering, and Weighted Prediction

Given a new input x, the population P is scanned for active or *matching* RFs, that is, all RFs with an activity $\phi(x) > \exp(-1)$, which relates the radius of an RF directly to the radius parameters of the Gaussian kernel function from Equations (3.7) to (3.9).

If no RF matches the input (e.g. because of the initially empty population), a new RF with random shape is generated, with its center at x. This is called *covering* and ideally does not occur when the population has reached its maximum size (see Algorithm 4.1).

Algorithm 4.1: Matching and covering.

    input : Population P of all RFs and input x
    output: Matchset M of active RFs

1   $M \leftarrow \emptyset$
2   **for** RF $\in P$ **do**
3   **if** RF matches x **then**   // most CPU time here
4   add RF to M

5   **if** M is empty **then**   // covering
6   create RF with center $c = x$, random shape and size
7   add RF to M

8   **return** M

Each of the matching RFs computes its local prediction $p_i(x)$ using the internal model – be it a constant, a linear, or even quadratic model. A weighted average

$$p(x) = \frac{\sum_i \varphi_i p_i(x)}{\sum_i \varphi_i}, \quad 1 \leq i \leq |M|, \tag{4.1}$$

where φ is the fitness of the i'th RF, which will be explained later. For now think of it as quality such that high quality RFs have more influence on the prediction (see Algorithm 4.2).

4.3 Local Model Adaptation

When the actual function value y at time t is known, the algorithm can improve its local, parametric models. Therefore the local prediction error $\|y - p_i(x)\|$ of the i'th RF is used to adapt the model in a gradient descent fashion. As discussed earlier, the Recursive Least Squares (RLS) method performs very well with a linear model [56].

Algorithm 4.2: Weighted Prediction.

input : Matchset M and input x
output: Predicted function value p

1 $p \leftarrow 0$
2 $s \leftarrow 0$ // sum of weights
3 **for** $\mathsf{RF}_i \in M$ **do**
4      $p_i \leftarrow \mathtt{predict}(\mathsf{RF}_i, x)$ // query individual model
5      $\varphi_i \leftarrow$ fitness of RF_i
6      $p \leftarrow p + \varphi_i p_i$ // fitness weighted contribution
7      $s \leftarrow s + \varphi_i$
8 $p \leftarrow p/s$ // normalize
9 **return** p

Additionally, each RF in the matchset M updates several internal variables required for the global structural evolution. The experience

$$\xi = \xi_{\text{old}} + 1 \qquad (4.2)$$

is the number of update calls so far. Next, the prediction error for the current RF is estimated. Two problems arise here: First, the error can only be computed from iterative samples which is inevitably inaccurate at the beginning. If the model is static, a running mean of the prediction errors would converge to the true value. However, the local models are dynamically changing with every new sample, which is the second problem for an accurate estimate of the prediction error. An optimal estimate could be computed when all samples are stored and, whenever the model changes, its prediction error is computed against all samples. This is a memory intense approach and typically infeasible for an infinite data stream. Instead, the prediction error estimate is updated by means of a modified delta rule

$$\varepsilon = \varepsilon_{\text{old}} + \max\left(\frac{1}{\xi}, \beta\right) \underbrace{(\|y - p_i(x)\|}_{\text{current error}} - \varepsilon_{\text{old}}), \qquad (4.3)$$

where β is the learning rate for well experienced RFs. The so called *moyene adaptive modifiée* update rule [90] applied here initially computes a moving average for t steps until $1/t < \beta$. From then on the delta rule is applied as is. The same update rule is used to estimate the matchset size

$$\psi = \psi_{\text{old}} + \max\left(\frac{1}{\xi}, \beta\right) (|M| - \psi_{\text{old}}), \qquad (4.4)$$

where $|M|$ denotes the number of RFs in M. The size of a matchset is dynamic as well, since location, shape, and size of RFs may change over time. Thus, the modified delta rule is appropriate here as well. The estimated matchset size is later used as a crowding criterion during deletion: A large ψ value indicates overcrowded regions of the input space, which are favored by the deletion mechanism explained later.

Finally, the evolutionary algorithm requires a fitness value, which should relate to accuracy as the accuracy should be maximized. However, the true accuracy of RFs is unknown, because only a finite number of samples is available. Furthermore, maximization of accuracy directly implies minimization of RF size, as smaller RFs have better accuracy unless the underlying data is trivial (fitted perfectly by the used model). Last, but not least, the GA should not focus on *simple* regions where high accuracy can be achieved, but the full sampling area should be covered uniformly. All together, the formula looks simple but implies a tricky balance between accuracy and crowding. The accuracy is computed relative to a target error ε_0 and cannot exceed one.

$$\kappa = \min\left(\left(\frac{\varepsilon_0}{\varepsilon}\right)^\nu, 1\right) \tag{4.5}$$

The updated fitness of a machting RF is given as

$$\varphi = \varphi_{\text{old}} + \beta\left(\frac{\kappa}{\sum_{i=1}^{|M|} \kappa_i} - \varphi_{\text{old}}\right), \tag{4.6}$$

where κ_i is the accuracy of the i'th RF in the current matchset. Thus, fitness is shared within the matchset according to relative accuracy. Put differently, each update offers a maximum, total fitness increment of one, reduced by learning rate β, shared among all RFs according to their relative accuracy. Accurate RFs receive a greater share of the fitness, even greater without overlapping competitors. Indefinite shrinkage is prevented by the accuracy cap of one. Algorithm 4.3 summarizes the update rules.

4.4 Global Structure Evolution

If the GA is run every iteration, the uniformity of coverage cannot be guaranteed: Certain areas might get more samples and therefore produce more offspring on GA application. To balance the coverage over the whole input space, the GA is only applied when the matchset has an average age greater than the threshold θ_{GA}. Therefore each RF stores the timestamp (iteration) of creation – due to covering or GA – and the average timestamp of the matchset is compared against the current iteration.

Algorithm 4.3: Local updates of matching RFs.

input : Matchset M, input x, true function value y, learning rate β

1   $\kappa_{\text{sum}} \leftarrow 0$ // sum of accuracies

2 **for** $\text{RF}_i \in M$ **do**

3   updateModel(RF_i, x, y)   // update individual model

4     $\xi_i \leftarrow \xi_i + 1$ // Eq. (4.2)

5     $\varepsilon_i \leftarrow \varepsilon_i + \max(1/\xi, \beta) \left(\| y - \text{predict}(\text{RF}_i, x) \| - \varepsilon_i \right)$ // Eq. (4.3)

6     $\psi_i \leftarrow \psi_i + \max(1/\xi, \beta) \left(|M| - \psi_i \right)$ // Eq. (4.4)

7     $\kappa_i \leftarrow \min((\varepsilon_0/\varepsilon_i)^\nu, 1)$ // Eq. (4.5)

8     $\kappa_{\text{sum}} \leftarrow \kappa_{\text{sum}} + \kappa_i$

9 **for** $\text{RF} \in M$ **do**   // κ_{sum} required for loop

10    $\varphi_i \leftarrow \varphi_i + \beta((\kappa_i/\kappa_{\text{sum}}) - \varphi_i)$ // Eq. (4.6)

On a mature matchset, the steady state GA can begin its work. Two RFs are selected from the matchset using probabilistic tournament selection with a relative tournament size of $0 < \tau \leq 1$. The number of selected RFs (two) is conservative, but an increased reproduction rate might hinder learning on complex problems [81]. In turn, the selected RFs are cloned and their fitness is reduced by 90%. A uniform crossover operator is applied with probability χ, mutation is applied individually to each allele with probability μ. The process is illustrated in Algorithm 4.4.

Algorithm 4.4: Selection, reproduction, crossover, and mutation.

input : Matchset M, relative tournament size τ, crossover
          probability χ, mutation probability μ

output: Generated offspring RF_1, RF_2

1 **for** $i = 1$ **to** 2 **do**

2     $T \leftarrow$ select τ percent randomly from M

3     $\text{RF}_{\text{best}} \leftarrow$ select the RF with maximum fitness from T

4     $\text{RF}_i \leftarrow$ Clone RF_{best} and reduce its fitness

5 crossover ($\text{RF}_1, \text{RF}_2, \chi$)

6 mutation (RF_1, μ)

7 mutation (RF_2, μ)

8 **return** RF_1, RF_2

The actual type of crossover and mutation operators mainly depends on the chosen kernel representation. The following section briefly illustrates useful operators for general ellipsoidal kernels as introduced in Section 3.1.1.

4.4.1 Uniform Crossover and Mutation

First, a uniform crossover operator is applied to two given kernels with probability χ, which is one of XCSF's parameters. A general ellipsoidal kernel is defined by its n-dimensional center, n radii, and $n(n-1)$ rotation angles as stated in Equation (3.9). Given two such kernels the uniform crossover routine exchanges a single allele between the two with 50% chance, where each element of the center and every radius represents an allele.

Unfortunately it is non-trivial to apply crossover to a higher-dimensional rotation because all angles jointly define the rotation and, thus, cannot be treated separately [82]. Thus, the crossover operator exchanges the whole rotation matrices with 50% chance. On the one hand, crossover serves as an exploration tool to combine features of *good* parents into new, possibly superior offspring. However, crossover also distributes good solutions among neighboring RFs which speeds up the learning process.

After application of crossover, a mutation operator is applied to each individual. First, the center of each kernel is mutated, where each allele is mutated with probability μ. With a smooth function and the notion of similarity in mind, mutation ideally acts as a local search. Therefore the center is shifted to a random location within the current geometric shape. For a general ellipsoid this can be done by first generating a random location within the unit sphere, then stretching this sphere by the radii of the ellipsoid, and finally rotating the axis-aligned ellipsoid with the current rotation matrix. This sounds fairly complex, but all elements are readily available and it boils down to the generation of an n-dimensional random vector and a matrix multiplication [82].

After the center has been shifted within the kernel, the radii are mutated, again each one with probability μ. Care has to be taken to avoid a bias towards enlargement or shrinkage. Therefore, each radius is either shrunk or stretched by a random factor in $[1, 2)$. This provides an unbiased, scale invariant, local search operator.

Unless an axis-aligned representation is chosen, the orientation of the kernel is mutated as well. Equation (3.9) specifies the kernel rotation via matrix V, which can be altered by multiplication with other rotation matrices generated from random angles as follows. First, $n(n-1)$ rotation angles are generated, one angle for each rotation plane formed by two axes. With probability μ an angle is taken uniformly random from $[-45°, +45°]$ and set to zero otherwise. Rotation by angle α about the plane formed by axes i and j (with $0 \leq i < j \leq n$) can be expressed as a matrix $\boldsymbol{G}(i, j, \alpha)$ with

$$
\begin{aligned}
G_{kk} &= 1 && \text{for } k \neq i, j \\
G_{ii} &= \cos(\alpha) && \text{and} \quad G_{jj} = \cos(\alpha) \\
G_{ij} &= \sin(\alpha) && \text{and} \quad G_{ji} = -\sin(\alpha),
\end{aligned}
\tag{4.7}
$$

where other entries are zero. Thus, the original rotation matrix $\boldsymbol{V}$ is multiplied with $n(n-1)$ such sparse rotation matrices. Since only few non-zero angles are generated and only two columns of the respective rotation matrix are different from a unit matrix, the overall computational cost is low.

4.4.2 Adding new Receptive Fields and Deletion

The new offspring are inserted in the existing population of RFs. What happens, when the maximum number of RFs is reached and the GA evolves some more RFs? In this case – that is the major learning phase – other RFs are removed from the population based on a heuristic vote v that respects crowding, experience, and accuracy

$$
v = \begin{cases} \psi \, \frac{\varphi_{\text{avg}}}{\varphi} & \text{if } \xi > \theta_{\text{del}} \wedge \varphi < 0.1 \, \varphi_{\text{avg}}, \\ \psi & \text{otherwise}, \end{cases}
\tag{4.8}
$$

where ξ is the experience from Equation (4.2), ψ is the estimated matchset size (Equation (4.4)), φ is the individual fitness of the RF in question (Equation (4.6)), φ_{avg} is the average fitness of the whole population, and θ_{del} is a threshold parameter that separates experienced from non-experienced RFs. When a RF is experienced but its fitness is less than $1/10$th of the average fitness (upper case) the deletion vote is boosted by the factor $\varphi_{\text{avg}}/\varphi > 10$. This is a strong pressure to remove badly performing RFs from the population. Apart from that, the deletion balances the coverage of the input space, as highly crowded regions will have a correspondingly greater estimate ψ. The deletion *probability* of a single RF

$$
\frac{v}{\sum_{i=1}^{|P|} v_i}
\tag{4.9}
$$

is its deletion vote relative to the votes of all other RFs. The deletion process can be realized by a roulette wheel selection as illustrated in Algorithm 4.5.

4.4.3 Summary

XCSF is a fairly complex algorithm due to multiple objectives inherent to the task of learning a suitable clustering of an unknown input space optimized for

---

Algorithm 4.5: Deletion via roulette wheel selection.

input : Population P of size N, number k of RFs to delete, deletion experience threshold θ_{del}

1   $\varphi_{\text{avg}} \leftarrow 0$
2   **for** RF $\in P$ **do**   // compute average fitness
3      $\varphi_i \leftarrow$ fitness of RF
4      $\varphi_{\text{avg}} \leftarrow \varphi_{\text{avg}} + \varphi_i / N$

5   $w \leftarrow$ vector of length N // roulette wheel
6   $w_0 \leftarrow 0$ // summand for $i = 1$
7   **for** $i \leftarrow 1$ **to** N **do**   // compute deletion votes
8      $\xi_i \leftarrow$ experience of RF$_i$
9      $\psi_i \leftarrow$ matchset size estimate of RF$_i$
10     $\varphi_i \leftarrow$ fitness of RF$_i$
11   **if** $\xi_i > \theta_{\text{del}}$ **and** $\varphi_i < 0.1\,\varphi_{\text{avg}}$ **then**   // Eq. (4.8)
12        $w_i \leftarrow w_{i-1} + \psi_i\,\varphi_{\text{avg}} / \varphi_i$
13   **else**
14        $w_i \leftarrow w_{i-1} + \psi_i$

15   **while** $k > 0$ **do**   // delete one
16     $r \leftarrow$ uniformly random from $[0, w_N)$
17     $i \leftarrow 1$
18   **while** $w_i \leq r$ **do**
19        $i \leftarrow i + 1$
20   remove i'th element from P and update w
21     $N \leftarrow N - 1$
22     $k \leftarrow k - 1$

---

an accurate model of a non-linear function surface with piecewise linear models. The whole process can be split up into three parts: Prediction or inference, micro optimization of local models, and the macro optimization of the kernel clustering.

When a population of RFs is to predict the function value at a given input x, all kernels compute their activity for that particular input. The active RFs compute individual predictions that are merged into a weighted sum, weighted by fitness, which is a kind of quality measure.

Micro optimization takes place at the level of local, typically linear, models. Each model minimizes the sum of squared errors in its respective subspace. Macro optimization instead works on the location, shape, and size of RFs by means of a GA. Evolution strives for a) RFs with low prediction error down

to a target error given as parameter ε_0, b) maximally general RFs that cover
as much space as possible while maintaining the target error, and c) a uniform
coverage of the whole input space.

4.5 Relevant Extensions to XCSF

XCS and its derivative XCSF are an active research field and various extensions
and variations were published. Some extensions deserve a section in this thesis
as they have proven to be generally useful, with only rare exceptions.

4.5.1 Subsumption

The GA pushes towards accurate RFs by means of the fitness pressure, while
highly general RFs naturally receive more reproduction opportunities during
selection. When the desired accuracy is reached, fitness is capped to a maximum value of one which essentially disables the fitness pressure beyond this
point. This enables the GA to not only evolve accurate, but also maximally
general RFs at the same time. The subsumption technique [97] further boosts
the generalization capabilities for accurate RFs.

When new offspring are inserted into the population, the population is first
scanned for experienced ($\xi \geq \theta_{sub}$) and accurate ($\varepsilon \leq \varepsilon_0$) RFs that eventually
cover the newly created child (Figure 4.2). The minimum experience θ_{sub} is
a parameter that should roughly reflect the number of samples required for a
local model to converge sufficiently. If the estimated prediction error ε of an RF
after θ_{sub} samples is below the target error ε_0, it is able to subsume less general
offspring. In that case, the child gets subsumed by the more general, yet still
accurate RF which essentially produces a clone of the more general one instead
of adding the small one.

Computationally, this can be simplified by augmenting all RFs with a so called
numerosity count that represents the number of copies – initialized to one. This
does not only save some memory, but also speeds up the matching process,
which requires most computational resources in XCSF – especially in higher
dimensions. The population size is computed as the sum of all numerosities
and thus, even in the case of subsumption, RFs have to be deleted before new
offspring can be inserted or subsumed.

Subsumption is best explained with an example. Suppose there is a optimal
RF in the population, where optimal means accurate (its prediction error is less
than the target error) and maximally general (increasing its size would result in
a prediction error greater than the target error). The GA reproduces this RF
and by chance produces a smaller sized RF. Moreover, suppose the parent covers
the offspring. The offspring is accurate as well, but overly specific. Instead of

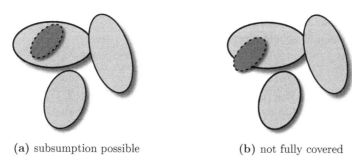

(a) subsumption possible   (b) not fully covered

Figure 4.2: A fourth RF (dashed contour) is inserted into the population. (a) The offspring is fully covered by another RF. (b) Here, the offspring covers a distinct part of the input space.

adding the overly specific RF to the population, the accurate and maximally general one is strengthened by incrementing its numerosity.

Given non-rotating geometric expressions (say rectangles or ellipsoids), the geometric processing is simple. However, with *rotating* kernels the test for inclusion becomes quite complex. The best way to compute the geometric inclusion is to use affine transformations and to combine them properly.

Suppose the RF A is to be inserted in the population and RF B wants to subsume A. Put differently: Does B contain A? A similar technique as for matching is used. In matching, a vector within the ellipsoidal coordinate system was mapped into a coordinate system where the ellipsoid is actually the unit

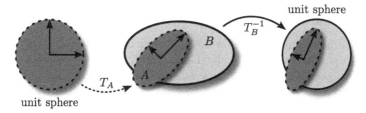

Figure 4.3: The affine transform T_A maps the unit sphere onto the general ellipsoid A (dashed contour). The inverse transform T_B^{-1} maps the ellipsoid B back onto the unit sphere (solid contour). Combining both transforms maps A onto B's unit sphere representation (right-hand side). Here, ellipsoid A is not fully contained in B and subsumption does not apply.

sphere (inverse affine transformation). Here, the whole ellipsoid A is mapped back into the unit sphere representation of B:

$$T = T_B^{-1} T_A, \qquad (4.10)$$

where T_A is the affine transformation of ellipsoid A and T_B^{-1} is the inverse transformation of ellipsoid B as illustrated in Figure 4.3. Next, both end points of each principal axis of the transformed ellipsoid A are tested for inclusion in the unit sphere, which is the case when those transformed vectors have a length less than one. Therefore all positive unit vectors with $e_i^{(+)} = 1$ and their negative counterparts $e_i^{(-)} = -1$ are transformed and their length is computed. The condition for geometric inclusion can be formulated as

$$\forall_{0 < i \leq n} \quad \| T e_i^{(+)} \| \leq 1 \wedge \| T e_i^{(-)} \| \leq 1 . \qquad (4.11)$$

Since T is an affine transformation including translation the sign of the unit vector makes a difference: Figure 4.3 shows that one end point of A is not contained in B.

In sum, a matrix-matrix multiplication, $2n$ matrix-vector multiplications, and $2n$ vector length computations are required, when each RF provides its affine forward and inverse transform. Taken into account the rarity of adding new RFs (due to θ_{GA}) and the positive effect of subsumption, the computational complexity is justified. Furthermore, experience has shown that matching new inputs governs the computational demand by far on problems with dimensions $n > 5$. Subsumption is applied throughout all experiments in this thesis.

4.5.2 Condensation and Compaction

Another notable plug-in for XCS or XCSF are condensation techniques that reduce the population size after learning. The GA in XCS requires a certain *overhead* in population size to work on. When the algorithm has reached a sufficient accuracy, the evolutionary overhead can be reduced. However, the removal of RFs involves a complex geometric problem: *How to avoid holes in the coverage?* It turns out that this is an almost intractable geometric problem and heuristics are required.

Condensation

Stewart Wilson, the inventor of XCS, introduced the condensation technique [97], where simply mutation and crossover are deactivated after the desired number of learning steps. However, RFs are still reproduced, but reinserted without modification. With subsumption in place, the effect is that selection takes place and the winner gets its numerosity increased by one. The deletion mechanism is still active and removes less fit individuals and purges overcrowded regions. This way highly fit RFs become strengthened and inaccurate ones probably deleted over time[1]. Condensation is a simple, conservative way of cleaning evolutionary overhead. However, it does not guarantee that coverage is maintained. One way to avoid *holes* in the population is the so called *closest classifier matching*.

Closest Classifier Matching

Instead of using the exact geometric shape of an RF (classifier) to define a hard cap to match or not, alternatively the smooth kernel activation function can be used. Instead of considering only those RFs above a fixed activation threshold, all RFs compute their activity and the most active ones (closest in terms of activity) are considered, so called *closest classifier matching* [18]. While this method does not *remove* gaps (RFs are not trained on a gap, so the function must be smooth in between for this to work well), it gives at least some solution to gaps that can always occur during condensation. Furthermore, it clearly relates to other kernel smoothing approaches such as RBFN or Locally Weighted Projection Regression (LWPR), where activity is used to weight different local models without a threshold.

Greedy Compaction

The closest classifier matching idea avoids holes in the coverage and therefore a more aggressive reduction of population size is possible. The so called *greedy compaction* algorithm [18] first selects all experienced RFs ($\xi > \theta_{\mathrm{sub}}$) from the population and sorts them according to the estimated prediction error. The most accurate RF A is considered first and all other RFs B that match its center get subsumed by A, that is, the numerosity of A is incremented by the number of RFs B, which are deleted. Then the next RF is treated analogously. This is more aggressive than the conservative condensation procedure described before and consequently reduces the population size much faster.

The subsumption and condensation extensions don't simplify the algorithmic trickery of XCSF. The following chapter aims to explain *why* and *how* XCSF works at all from a macroscopic viewpoint. Armed with a better understanding of the system the *challenges* for XCSF and possible *enhancements* are discussed in detail. Those two chapters make up the analytic part of this thesis.

---

[1]Technically this does not reduce the population size but the number of distinct RFs.

Part II
Analysis and Enhancements of XCSF

5 How and Why XCSF works

While the previous chapter gave details of the mathematical description of XCSF, this chapter sheds light on the resulting behavior. XCS, the big brother of XCSF, has already been analyzed thoroughly. While a large part of the theory also applies to XCSF, there are differences that require a different viewpoint. The present chapter reviews the information relevant for XCSF from [16, 17, 13] with the goal to illustrate *how* and *why* the algorithm works.

The first section reviews the partially conflicting goals of XCSF and how they can be achieved. The balance between four identified objectives is discussed in Sections 5.2 and 5.3. Finally, Section 5.4 illustrates the three typical phases of the algorithm.

5.1 XCSF's Objectives

Accuracy is the obvious target, however, infinite accuracy may cost infinite resources (cf. Cybenko theorem [23]), which is impractical. Therefore XCSF has a target error ε_0 to reach. Higher accuracy generally means smaller receptive fields.

Coverage of the full input space is another simple requirement. With a certain accuracy, that is, small receptive fields, the number of required Receptive Fields (RFs) to cover the full space has a lower bound. Furthermore, overlap is almost inevitable on a real-valued environment, but reduces the effective coverage.

Certainly, those two goals can be met with an large number of receptive fields, but an optimal representation also has high effectiveness (low computational cost). The next two goals relate to resource management.

Generality refers to the ability to make predictions on unknown inputs due to knowledge about closely related inputs. Thus, generalization is an important feature of Machine Learning (ML). In the context of XCSF, generality directly relates to RF volume and it is desirable to have large RFs that are still accurate.

Low Overlap refers to an efficient distribution of RFs. Every region of the input space should be covered with a similar number of RFs such that resources are distributed uniformly.

While the first two goals, *accuracy* and *coverage* increase the required population size, the latter two opt for a reduced population size – thus, they need to be balanced. Put differently, Accuracy and generality are conflicting goals. There is a rather clear balancing between accuracy and generality via the target error, which defines optimality with respect to those two goals. Given those are satisfied, coverage is a lower bound on the population size and low overlap essentially says that this lower bound should not be exceeded. So, how are the goals achieved?

5.2 Accuracy versus Generality

Accuracy and generality are balanced by the desired target error that is set by the user. Receptive fields need to shrink until they satisfy the target accuracy but then expand to cover as much space as possible without increasing the prediction error beyond the target. A receptive field that is accurate but cannot expand any further without being inaccurate is said to be an *accurate, maximally general* receptive field. How is this balance realized in XCSF[1]?

Fitness Pressure. Accuracy (Equation (4.5)) is directly encoded into fitness (Equation (4.6)). Consequently the Genetic Algorithm (GA) searches for accurate RFs in the first place – so called fitness pressure. However, once the prediction error reaches the target error, accuracy is capped to a maximum of one. Therefore the GA has no further incentive to produce individuals with higher accuracy (see Figure 5.1).

Subsumption Pressure. Once, a receptive field is accurate, it is able to subsume smaller (more specific) RFs as detailed in Section 4.5.1. Thus, subsumption puts a pressure in the opposite direction of fitness pressure – these disjoint pressures meet at the point of optimality.

Set Pressure. In addition to fitness and subsumption, the selection mechanism of the GA implicitly introduces another pressure. Large RFs cover a greater fraction of the input space and, thus, the probability to be in a matchset is larger as well (actually proportional to their volume, as long the input space boundaries are not exceeded). Compared to smaller RFs with the same

---

[1]An in-depth analysis of this balance in XCS can be found in [13, Chapter 5]

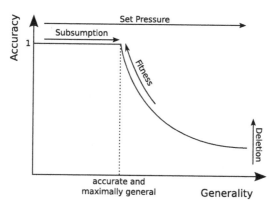

Figure 5.1: Several pressures work either for accuracy or generalization. Accuracy (defined in Equation (4.5)) increases polynomially in ν until the target accuracy is achieved, where it becomes flat. Once accurate, subsumption pushes towards generality. Selection offers higher chances to large RFs for their higher matching probability (Set Pressure). Deletion strongly penalizes below average fitness (initially) but also high generality (fine tuning).

fitness, those large ones have a better chance to reproduce. Thus, set pressure fosters generalization.

Deletion Pressure. If the GA evolves a new RF but the maximum population size is already reached, another RF has to be deleted. The probability is based on the estimated matchset size (to reduce crowding) and the relative fitness as detailed in Equation (4.8). Primarily, deletion removes RFs below average in terms of fitness – if their fitness is less than 1/10th of the average fitness of the population, the vote increases by a factor 10 as illustrated in Figure 5.2.

Second priority is given to crowding, where the probability increases linearly with the estimated matchset size ψ. This leads to the balancing of coverage and overlap.

5.3 Coverage and Overlap

With accuracy and generality properly balanced, another important step is to distribute the RFs such that a) the full input space is covered, but b) without too much overlap. Initially, XCSF's population is empty and new samples are covered by randomly generated RFs. But once the population is full, covering would require other RFs to be deleted. Eventually, another hole may appear by deletion and covering would occur again. The so called *covering-deletion*

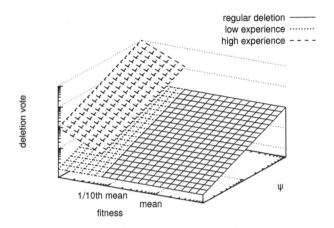

Figure 5.2: A log-log-log plot of the deletion formula. Below 1/10th of the average fitness the penalty jumps by a factor of 10 for experienced RFs.

cycle [16] must be prevented by a suitable population size that is large enough to not only hold as many RFs to fully cover the input space, but also with some extra space for the GA to *work on*. The user is responsible to set the population limit according to the problem at hand.

Finally, a reduced overlap is the last of the goals and it also has the least priority. Looking at Equation (4.6), the fitness is shared among the whole matchset. Consequently, overcrowded regions will receive less fitness per RF, even if they are accurate and a lower fitness implies reduced reproductive opportunities. Apart from the fitness sharing, the GA is balanced to occur only when a matchset has a certain age. Moreover, the deletion favors overcrowded regions (see Figure 5.2). Altogether, these three mechanisms help to cover the whole input space roughly uniformly and, thus, distribute XCSF's resources suitably. In addition a condensation technique [97, 18] can be applied after sufficient learning to further purge the evolutionary overhead as described in Section 4.5.2.

5.4 Three Phases to Meet the Objectives

XCSF's learning can be separated in at least two, eventually three different phases. First, the input space must be covered by randomly generated RFs. When the population size approaches the limit, the deletion mechanism begins to remove badly performing RFs and therefore further pushes the population

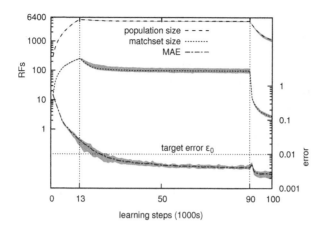

Figure 5.3: A typical performance graph for XCSF showing population size, the matchset size, and Mean Absolute Error (MAE). The shaded area represents minimum to maximum performance of 20 independent runs, while the lines are median performance. At about 13 k iterations, the population is full and RFs must be deleted to make room for new offspring. The target error is reached at about 24 k iterations. Condensation commences at 90 k iterations and reduces the population size by roughly 80%.

towards an accurate solution. Optionally, condensation reduces the population size after learning. Figure 5.3 shows a typical performance graph for XCSF to illustrate the different stages.

Initial Phase: Covering and Reproduction. When no RF is responsible for a new input, a random one is generated. Typically, the size of initial RFs is chosen comparably large, such that the input space is quickly covered and the GA can start to work. If the initial size is small, the population quickly fills with small random RFs and evolutionary selection is unable to work, as only few RFs cover a certain input and, thus, the selection is among those few individuals.

Training Phase: Deletion stresses Accuracy. Once the maximum population size is reached, deletion further increases the pressure towards accurate RFs. Now, over-sized RFs are removed which is also visible at the average size of the matchsets, which drops slightly. This is the major learning stage where XCSF should be able to approach the target error set by the user, unless the problem is too difficult, the target error is set too low, or the population size is chosen too small for the problem at hand.

Covering should not happen anymore, since it now typically means a loss of knowledge: A completely random RF is generated, but another RF has to be removed from the population – eventually an experienced, well shaped one. Instead, the population size should be large enough such that the full input space can be covered with randomly generated RFs. Alternatively, the initial volume of covering RFs can be increased, such that fewer RFs suffice to cover the full space. On some problems, however, the GA may take a while to find a suitable clustering of the input space and covering may still happen until sufficiently general RFs are evolved such that more resources are available for other parts of the input space.

Condensation Phase: Evolutionary Cleanup. When the accuracy is sufficient, the GA can be deactivated and the evolutionary overhead purged. Typically this slightly reduces the accuracy, but interestingly the approximation accuracy may even improve (see Figure 5.3). If the optimal solution is a small niche in feature space, the mutation and crossover operators produce worse RFs with a high probability once a good solution was found. Those badly performing RFs are removed by condensation and, in turn, the accuracy increases even further. However, if the problem at hand requires most of the available resources, that is, most of the available RFs to accurately model the function surface, the effect is negative and approximation performance may reduce slightly.

The present chapter briefly illustrated how XCSF works and described the balance between accuracy and generality, which is a key point in the algorithm. The next chapter discusses challenging problems for XCSF.

6 Evolutionary Challenges for XCSF

Training local models is a comparably simple task. Optimizing the kernel shape such that local models fit well to the underlying data, on the other hand, can be a demanding challenge. In XCSF, the Genetic Algorithm (GA) is responsible for this optimization and the subtle mechanisms in the fitness estimation guide the GA towards a balance between accuracy and generalization. This chapter is concerned with *challenging* problems, where XCSF may not find an accurate solution out of the box.

The term *challenging* may refer to different issues: An algorithm may require huge amounts of memory to solve a certain problem properly. When enough memory is available, a challenging problem may still require a large number of learning iterations. Eventually, a problem cannot be solved at all because available hardware is not sufficient or the algorithm is not suited for the problem. Moreover, the *user* may find it challenging to obtain good parameters for an algorithm such that a given problem can be solved at all – parameter tuning is not the topic here.

The memory requirements of XCSF mainly depend on the population size, that is, the number of Receptive Fields (RFs) required for a given problem. Section 6.1 analyzes XCSF's *scalability* with respect to population size in a tractable, worst case scenario. The mathematical relation between population size, target accuracy, and problem dimension is worked out. Furthermore, it is shown that a suitable representation can improve the scalability by orders of magnitude.

The *learning time*, in terms of iterations, is often closely tied to the above mentioned factors, because the optimization of a larger population also takes more iterations. Learning time reflects the required effort of the GA to find a suitable kernel structure and Section 6.2 illustrates one way to reduce this learning time, especially in high-dimensional problems. Therefore XCSF is empowered with a new *informed* mutation operator that incorporates knowledge about recent samples, while usually a blind, completely random mutation is applied. Thus, the evolutionary search is *guided* towards well-suited RFs.

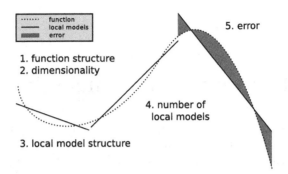

Figure 6.1: Five elements are relevant for XCSF. The given problem has a certain structure (e.g. polynomial) and a certain number of relevant dimensions. On XCSF's side two elements can be tuned: The type of local model (here linear) and the number of local models (three). The resulting approximation error further depends on the clustering evolved by the GA which is assumed to be optimal for now.

6.1 Resource Management and Scalability

Memory requirements of algorithms such as XCSF, Locally Weighted Projection Regression (LWPR), or Radial Basis Function Networks (RBFNs) are governed by the number of local models required to successfully approximate a given problem. However, four more elements are deeply intertwined with the number of local models: the function structure, its dimensionality, the chosen type of local model, and the approximation accuracy (Figure 6.1). The present section tries to disentangle these five factors to shed some light on the scalability of kernel regression algorithms. Therefore the *optimal* population size is derived for a particular set of functions and empirical experiments reveal the true scaling of XCSF. The experiments are done with XCSF, but the general message applies to other kernel regression algorithms, including RBFN and LWPR as well[1].

6.1.1 A Simple Scenario

The general idea is to start with a single model and derive the mathematical relation between function properties, model type, and resulting approximation error. In turn, the formula is generalized from a single model to a population of local models. The major problem is that the approximation error is not graspable in a general setup, because for equally sized RFs the error at different locations in input space is not necessarily the same. Both the function *structure*

---

[1]Section 6.1 is largely based on [86].

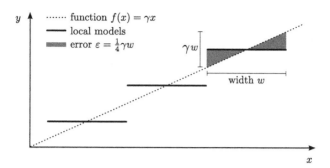

Figure 6.2: Three non-overlapping, constant models approximate a linear function $f(x) = \gamma x$. All RFs have the same width w resulting in the same approximation error ε (shaded area) that solely depends on the width w of a RF and the gradient γ of the function.

and local model *structure* account for the error in a typically non-linear fashion. A deeper analysis is only possible in a simplified scenario.

Assuming a linear function and locally constant models that do not overlap results in a tractable scenario as illustrated in Figure 6.2. Here, the function structure is clearly defined as the gradient, which can be used to compute the prediction error of a particular RF.

Let

$$f : [0,1]^n \to \mathbb{R}$$

$$f_\gamma(x) = \gamma \sum_{i=1}^{n} x_i \tag{6.1}$$

be a parametrized, linear, n-dimensional function, where γ defines the gradient. When XCSF is equipped with locally constant models, the gradient γ directly defines the difficulty level. Without loss of generality, inputs are taken from the $[0,1]^n$ interval and consequently the volume of the input space is one, which simplifies later equations. The function structure is uniform, that is, the same RF size yields the same error independent of the actual location of the input space.

First, suppose a single constant model covers the full input space $[0,1]^n$, that is, an RF with volume one. The approximation error depends on the gradient γ and dimension n as depicted in Figure 6.3. Increasing γ results in a linear increase of the corresponding error ε. The approximation error is also influenced *linearly* by the RF width w (cf. Figure 6.2) and can be written as

$$\varepsilon = \Gamma w \,, \tag{6.2}$$

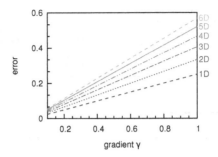

Figure 6.3: The approximation error of an RF that covers the $[0, 1]^n$ interval for dimensions $n = 1, \ldots, 6$. The horizontal axis shows the gradient, which has a linear influence on the resulting error. For dimensions $n \geq 3$ the error was approximated with Monte-Carlo integration.

where Γ is some constant depending on γ, e.g. $\Gamma = 0.25\gamma$ in the one-dimensional case. The exact value of Γ is not of interest, but the emphasis is on the *linear* relation between error, width, and Γ. The latter variable Γ serves as a linear abstraction of *function difficulty*: Doubling the difficulty Γ results in twice the approximation error.

In the n-dimensional case, an RF's volume is $V = w^n$. Expressing the above relationship with respect to volume results in the following relation

$$\varepsilon = \Gamma w$$
$$= \Gamma \sqrt[n]{V}, \tag{6.3}$$

where the error still has a linear relation to the difficulty Γ.

6.1.2 Scalability Theory

A linear relation between *approximation error* ε and *function difficulty* Γ as depicted in Equation (6.3) forms the basis of the following theory that connects the problem dimensionality n, RF volume V, and population size N.

The *optimal* volume V_{opt} for a RF can be computed with respect to a certain target error ε_0. Optimal means that the RF is accurate, thus its error ε is below ε_0 and its volume is maximal while satisfying the former constraint. Put differently, the volume is maximized subject to the constraint $\varepsilon \leq \varepsilon_0$, which is the case for $\varepsilon = \varepsilon_0$. Substitution in Equation (6.3) and solving for the volume gives

$$V_{\text{opt}} = \left(\frac{\varepsilon_0}{\Gamma} \right)^n . \tag{6.4}$$

Thus, the optimal volume scales exponentially in the dimension and polynomially in the desired target error and difficulty.

The optimal *coverage* of the input space would be a non-overlapping patchwork of such optimal and rectangular RFs. Since the input space was confined to a volume of one, the minimal (optimal) number of RFs to accurately approximate a function can be written as

$$N_{\mathrm{opt}} = \left(\frac{\Gamma}{\varepsilon_0}\right)^n, \tag{6.5}$$

which is the main result of the present section. Thus, the population size scales *exponentially* in the number of relevant dimensions n. Moreover, the desired target error has a *polynomial* influence on the population size. Finally, if the function complexity is doubled – that is, the absolute approximation error is doubled – then the population size needs to increase by a factor of 2^n to compensate for the increased difficulty.

6.1.3 Discussion

The assumptions made to get here were severe and certainly a linear function is not interesting at all, nor is a constant model interesting. The assumptions are briefly discussed below.

First, the function structure and data distribution were assumed to be uniform such that the approximation error does not depend on the location of the input space. For real-world data this does usually not hold, but can be seen as a worst case: The function is equally difficult *everywhere*.

Next, a linear function was tested against constant predictors – actually this is not required. Instead, the critical assumption is that the relationship between prediction error, RF width, and difficulty is *linear*. However, for non-linear functions and suitable prediction models the definition of this particular "linearly increasing difficulty" is non trivial. Nonetheless, a linear function *is* challenging for XCSF, when only constant predictors are applied.

Third, the population was taken to be *optimal*, that is, the input space is covered by the minimal number of accurate RFs without overlap. Therefore, a rectangular RF shape was required. Several mechanisms in XCSF aim for an optimal coverage, but this is generally not achievable on a continuous input space, especially when non-rectangular RF structures such as ellipsoids are chosen. How strong the last assumption is violated and if the scalability theory (Section 6.1.2) is true at all can be tested empirically.

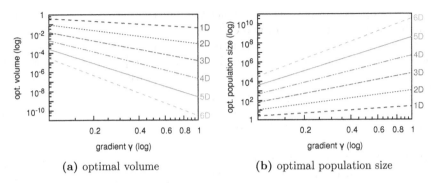

(a) optimal volume   **(b)** optimal population size

Figure 6.4: Illustration of Equations (6.4) and (6.5) for a target error of $\varepsilon_0 = 0.01$ on the linear function f_γ. The gradient γ can be seen as the difficulty level, which is shown on the horizontal axis. (a) In theory, the optimal volume decreases polynomially with increasing gradient which shows up as a straight line on a log-log plot. (b) The corresponding population size would increase polynomially in the dimension n.

6.1.4 Empirical Validation

Let us recall the basic setup of the simple scenario from Section 6.1.1. A linear, n-dimensional function $f_\gamma(x_1, \ldots, x_n) = \gamma \sum_{i=1}^{n} x_i$ was sampled uniformly random in the $[0, 1]^n$ interval. XCSF is required to reach an accuracy of $\varepsilon_0 = 0.01$ with constant predictors and rectangular shaped RFs as illustrated in Figure 6.2. The resulting approximation error of a single RF that covers the full input space was depicted in Figure 6.3.

This data is plugged into Equations (6.4) and (6.5), which results in *theoretically optimal* volume and population size with respect to the gradient γ as shown in Figure 6.4. Both the horizontal and vertical axis are scaled logarithmically and, thus, the polynomial functions are drawn as straight lines which allows for a good comparison between different dimensions.

In order to confirm the theoretical findings, XCSF's population size and RF volume is analyzed on the given linear function with varying gradient in dimensions one up to six. The learning time is set sufficiently large (500 k iterations) such that the GA is able to evolve a reasonable clustering of the input space. Condensation is applied during the last 10% of the learning time to reduce the evolutionary overhead in population size[2].

The population size in XCSF is fixed by a parameter but does not reflect the optimal number of RFs. Here instead a so called *bisection* algorithm searches in

---

[2] Other parameters were set to default values: Initial RF width is taken uniformly random from $[0.1, 1)$. $\beta = 0.1$, $\nu = 5$, $\chi = 1$, $\mu = 0.05$, $\tau = 0.4$, $\theta_{GA} = 50$, $\theta_{del} = 20$, $\theta_{sub} = 20$. GA subsumption was applied. Uniform crossover was applied.

a binary fashion for the minimal population size required to accurately approximate a certain function f_γ. The bisection algorithm runs XCSF with different population sizes until a reasonable size is found where the final population is accurate with respect to the given $\varepsilon_0 = 0.01$ error threshold. Three values are measured for each setting found this way: the number of RFs during learning, that is, before condensation, the final number of RFs after condensation, and the average RF volume after condensation. More details about this procedure can be found in [86].

The results are displayed in Figure 6.5, where again both axes are scaled logarithmically. This allows to compare polynomials as they show up as straight lines, where the gradient of the line equals the exponent of the original polynomial. The plots reveal that the empirical data is almost perfectly parallel to the developed theory, which confirms that the polynomial order fits reality. Apart from that, there is a slight gap between final population size and the theoretical minimum due to overlapping RFs (a factor of 1.90 ± 0.38 for all dimensions). Since the GA requires a substantial workspace to evaluate different shapes and sizes of RFs, the population size during learning is about a factor of 17.57 ± 2.39 larger than the optimum.

The RF volume on the other hand tends to become larger for higher dimensions. Overlapping RFs jointly predict a function value and, thus, their individual errors are typically greater than the prediction error of the final prediction, which explains why XCSF is able to reach the desired accuracy with even larger RFs than the developed equation predicts. Furthermore, XCSF's final prediction is governed by *accurate* RFs, but the volume is averaged over *all* individuals in the population. As this includes over-sized, low-fitness individuals the average volume is slightly biased. This bias increases with higher dimension n, as it is more difficult for the GA to approach the optimal RF structure. Nonetheless, the trend fits the developed theory well.

The results confirm the developed scalability model, which once again illustrates the curse of dimensionality: Generally, the population size grows exponentially in the problem dimension. More importantly however, the experiments have shown that XCSF uses its resources almost optimally. Thus, Equation (6.5) allows to better judge XCSF's performance on increasingly complex problems and also helps to set population size and target error accordingly.

6.1.5 Structure Alignment Reduces Problem Complexity

Checkerboard problems [88] are a worst case scenario for kernel regression techniques such as RBFN, XCSF, or LWPR: The approximation difficulty is the same in every dimension, at every spot of the input space and an optimal clus-

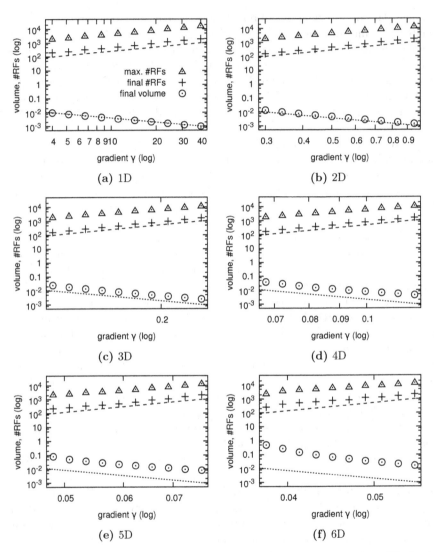

Figure 6.5: Results of the bisection procedure for dimension $n = 1, \ldots, 6$. The horizontal range is chosen such that the vertical range is the same for all dimensions which allows for good comparability. The theoretically optimal bounds from Figure 6.4 are drawn as lines, showing that the developed theory fits the real system quite well: The gradient of theory and data is roughly the same.

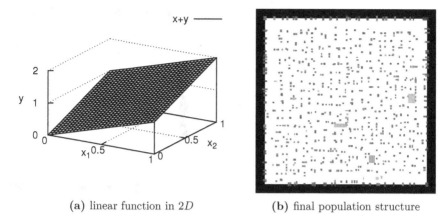

(a) linear function in $2D$ (b) final population structure

Figure 6.6: XCSF faces a checkerboard problem on linear functions, when equipped with rectangular RFs and constant local models. (a) The two-dimensional, linear function f_γ with $\gamma = 1$. (b) The final population structure is a patchwork of small squared RFs (shown at 20% of their actual size).

tering for such a problem looks like a checkerboard, that is, equally sized hyper cubes uniformly distributed over the input space as shown in Figure 6.6.

The problem is that the axis-parallel rectangular RF structure cannot express the shape of the constant subspaces of that function, that lie oblique in the input space. When a rotating RF structure is used instead, the problem can be learned with fewer resources because XCSF is able to cluster the input space more effectively. While the two-dimensional function initially represented a quadratic problem in terms of population size, a rotating RF structure reduces the complexity to linear as confirmed experimentally in Figure 6.7.

6.1.6 Summary and Conclusion

The memory demand of kernel regression algorithms such as RBFN, XCSF, or LWPR mainly depends on the number of RFs required to accurately approximate a given function surface. However, the optimal number of RFs is generally unknown and depends on multiple factors including the function structure itself, the employed RF structure, and the desired target error.

To shed some light on the memory requirements, a scalability model was developed and validated empirically in a worst case scenario. It was shown that the number of RFs inevitably scales exponentially in the number of dimensions, while the desired target error has a polynomial influence. The exponential influence of the problem dimension confirms the curse of dimensionality not

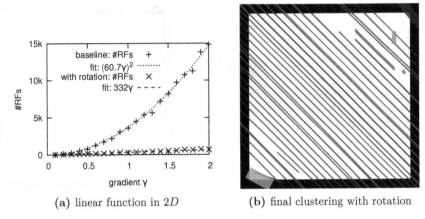

(a) linear function in $2D$ (b) final clustering with rotation

Figure 6.7: An appropriate RF representation enables XCSF to solve problems with fewer resources. (a) Comparison of the required population size with axis-aligned and rotating rectangular RF shapes on a two-dimensional, linear function. The data points are fitted with a respectively quadratic and linear function. (b) Rotation enables XCSF to exploit the constant subspaces in parallel to the $x_1 - x_2$ direction.

only for XCSF, but also for RBFN and LWPR, which both share the same underlying RF mechanism.

This provides an important opportunity: A problem is a worst case problem *only*, when the applied RF structure cannot express problem sub-structures. Put differently, when a suitable RF structure is used, even high dimensional input spaces may be clustered with a reasonable number of well-shaped RFs. As an illustrative example, the exponential worse case problem was turned into a linear problem by allowing the *rotation* of RFs.

6.2 Guided Mutation to Reduce Learning Time

In the previous section the memory requirements and the scalability to higher dimensions was analyzed and it was shown that a problem-dependent RF structure can reduce the memory requirements drastically. Nonetheless, the required *learning time* also increases with problem dimension, because the GA must optimize a larger number of RF parameters[3]. The present section illustrates how to endow the GA with local, informed optimization, which greatly reduces learning time on high dimensional problems[4]. An algorithm that combines global (evo-

---

[3] An exception are RF representations with constant complexity, independent of the dimension, such as spheres or cubes.
[4] Section 6.2 is largely based on [82].

lutionary) search and local (gradient descent) search is often termed *memetic algorithm* [38, 62].

The learning time in XCSF mainly depends on three facts: a) the chances to discover better structures in terms of fitness, b) the number of mutations from initial to optimal RF structure, and c) the number of RFs to optimize.

Regarding c), this depends on the problem at hand and the required population size, respectively. Furthermore, the problem structure plays an important role: If the optimal solution is roughly the same at every location in the input space due to a uniformly structured problem, then the crossover operator of the GA quickly spreads a good solution throughout the input space. On the other hand, if many different structures are required to accurately model the function surface, the crossover operator has less effect but may still help to find suitable structures for neighboring areas. In any case it depends on the problem and little improvement is possible here.

With respect to b), the minimal number of mutations from initial RFs to an optimal representation depends on the actual initialization (size, shape, rotation) and the step size of the mutation operator. Thus, the RF initialization parameters could be tuned for each problem anew, but still it depends on the problem and it should actually be the GA that finds optimal parameters, not the user.

Left is point a), which refers to the *chances* of improving RF fitness via mutation. Usually the mutation operator acts as a random local search and with increasing dimensionality of the evolutionary search space the chances to actually improve fitness become rather small. This is most apparent when rotating RF structures are applied which makes the dimension of the evolutionary search space quadratic in the problem dimension. Furthermore, rotation in dimensions $n \geq 4$ is not unique, meaning that different rotations of a RF can result in the same geometric shape. Rotation redundancy further bloats the search space of the GA.

However, the mutation operator need not solely rely on random search. A guided operator is developed for general ellipsoidal RFs in the following section.

6.2.1 Guiding the Mutation

General ellipsoids are a powerful representation (cf. Equation (3.9) in Section 3.1.1) that is well-suited for function approximation, especially when the relevant problem dimensions are not aligned to the input space coordinate system. Furthermore, the method described here fits best to ellipsoidal representations.

In n dimensions a general ellipsoidal shape is defined by n parameters for location, n radii, and $n(n-1)/2$ rotation angles and, thus, the evolutionary

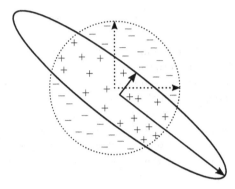

Figure 6.8: A RF (dashed) receives several samples during its lifetime for which its model is either accurate (plus symbols) or inaccurate (minus symbols). An ideal mutation would modify the shape such that only the accurate samples are covered (solid line), which results in an increased fitness.

search space has a substantial $n(n + 3)/2$ dimensions (one dimension for each allele that can be mutated). Additionally, the rotation angles are intertwined and modifying a single angle may not necessarily allow for an improvement in fitness. On the other hand, a lot of samples are given to individual RFs before mutation – but samples are not stored in XCSF for the large impact on memory. If a history of samples were available, that history could be used to *guess* good directions of mutation instead of randomly picking single alleles as sketched in Figure 6.8.

Certainly, not all samples can be stored on an infinite stream of data due to limited memory and processing capabilities. Thus, let K be the number of samples (x, y) stored for each RF. The first ingredient required for guidance is a *quality assessment* that tells which samples shall be covered and which ones should lie outside of an RF. Therefore a heuristic weight is computed for the k-th sample as

$$w_k = \left(\frac{\varepsilon_0}{|p(x_k) - y_k|} \right)^2 , \qquad (6.6)$$

where ε_0 is the desired target error and the denominator refers to the absolute error. If the absolute error equals the target error, w_k equates to one. If the error for one sample is large, the weight will be low and vice versa. The quadratic exponent puts an extra pressure into this formula – the actual value two, however, has no particular justification, but was found to be useful experimentally.

The next ingredient is that a general ellipsoidal shape can be defined by an (inverse) covariance matrix. From a statistics viewpoint a multivariate normal distribution forms an ellipsoidal shape. The inverse covariance matrix Q^{-1} of such a distribution can be plugged into Equation (3.9) as a distance metric. Then the ellipsoidal surface represents one standard deviation of the multivariate normal distribution.

If the covariance matrix of the stored K samples is computed and the samples are spherically distributed around the center of the RF, then the shape of the resulting covariance matrix roughly equals the original shape of the RF. This is not true for the *volume* (absolute axis length), but only for the *shape* (axis ratio), because the samples stored in a RF are not necessarily normally distributed. However, as long as the samples are distributed *spherically* around the center (e.g. uniformly in the input space), the shape can be well interpreted.

If a *weighted* covariance matrix is computed using the weighting according to Equation (6.6) instead of the regular covariance matrix, a low weight reduces the radius of the corresponding ellipsoid in that direction. On the other hand, a large weight increases the chances that the ellipsoid will actually cover that particular spot. The weighted covariance matrix is computed in two steps. First, the weighted mean of the samples x_k is computed as

$$m = \frac{\sum_{k=1}^{K} w_k x_k}{\sum_{k=1}^{K} w_k}. \tag{6.7}$$

Second, an individual entry of the weighted covariance matrix Q is given by

$$Q_{ij} = \Lambda \sum_{k=1}^{K} w_k (x_{k_i} - m_i)(x_{k_j} - m_j), \tag{6.8}$$

where Λ is a normalization factor. Usually a covariance matrix is normalized by the number of samples. However, this only affects the volume, which cannot be interpreted anyways. Instead, the volume is scaled such that the volume of the new, weighted covariance matrix Q equals the volume of the former parent RF. Now the inverse matrix Q^{-1} can be used as the distance metric for an offspring RF with its new center at the weighted mean m. The proper scaling of the volume can be efficiently combined with an inversion via Singular Value Decomposition (SVD) [34, 66] or Eigendecomposition [35, 66] – both are applicable to symmetric, positive semi-definite matrices.

Therefore let $VPV^T = Q$ be the decomposition of the unscaled covariance matrix, that is, for $\Lambda = 1$ in Equation (6.8). The diagonal matrix P contains the squared radii σ_i for the corresponding ellipsoid[5]. The volume of the n-dimensional ellipsoid is given as

$$\frac{2^{n-1}\pi}{n} \prod_{i=1}^{n} \sigma_i . \tag{6.9}$$

Given the actual volume v_a and the target volume v_t, the radii are scaled by the factor $\sqrt[n]{v_t/v_a}$. Consequently, the volume of the ellipsoid is not modified by the operator. The inversion is then realized by inverting the scaled radii and rearranging them into the inverse, diagonal matrix D with

$$D_{ii} = \frac{1}{\sqrt[n]{v_t/v_a}\sigma_i} . \tag{6.10}$$

Recombination of the scaled, inverted diagonal matrix D and the rotation matrix V yields a valid distance metric $sQ^{-1} = VDV^T$ with the same volume as the former ellipsoid but with new principal axes according to the prediction quality for the samples. The matrix VDV^T can be used in Equation (3.9) as a distance metric. The algorithmic description is summarized in Algorithm 6.1.

Algorithm 6.1: Guided mutation.

    input : n-dimensional RF with K samples x_k, y_k
1 compute weights w_k for all samples (x_k, y_k) from Equation (6.6)
2 compute weighted mean m from Equation (6.7)
3 compute unscaled, weighted covariance matrix Q from Equation (6.8)
4 $a_1 = \prod_{i=1}^{n} r_i$ // old radii to preserve volume
5 $VPV^T = \text{eigen}(Q)$
6 $a_2 = \prod_{i=1}^{n} \sqrt{P_{ii}}$ // new radii of unscaled Q
7 Create new $n \times n$ matrix D
8 **for** $i = 1$ **to** n **do**
9      $r_i = \sqrt[n]{a_1/a_2}\sqrt{P_{ii}}$ // rescale radii
10     $D_{ii} = 1/r_i^2$ // inverted, scaled diagonal matrix
11 set new distance metric to VDV^T
12 set new center to m

To sum up, if all weights were equal, the shape of the original RF would be reproduced. If the weights instead represent the *quality* of the RF, the shape is altered such that accurate areas are extended and the shape is shrunk in directions of large approximation errors as previously illustrated in Figure 6.8.

---

[5]Note that the corresponding matrix Σ from Equation (3.9) instead contains the *inverse* (squared) radii. Thus, an inversion is required.

Since the covariance matrix Q is computed from samples that are *not* normally distributed, the volume can not be interpreted and is instead scaled such that it equals the volume of the former RF. The scaled inverse of Q is then used as the distance metric of a new offspring.

It is important to see this operator as a heuristic, but not an optimal gradient descent. The quality criterion is chosen arbitrarily and the samples are not normally distributed. Furthermore, the operator becomes random if too few samples are available. At worst, the offspring has a zero volume because the samples form a hyperplane in the input space (e.g. $K = 2$). Consequently, the guided mutation operator should *support* the GA now and then, but the fine tuning of radii and volume is left to the regular mutation operator of the GA. Therefore the guided mutation is applied with a certain probability and otherwise regular mutation is applied. If too few samples are stored in one RF or the resulting covariance matrix is ill-conditioned the regular mutation operator is applied. The following section validates the approach on three complex Function Approximation (FA) problems of dimensions six and ten.

6.2.2 Experimental Validation

Consider the following n-dimensional, sinusoidal function

$$f_1(\boldsymbol{x}) = \sin\left(\frac{4\pi}{n} \sum_{i=1}^{n} x_i\right), \tag{6.11}$$

which is comprised of four periods in the $[-1, 1]^n$ interval. Another benchmark function is a modified version of the so called crossed ridge function [91]

$$f_2(a, b) = \max\left\{\exp(-10a^2), \exp(-50b^2), 1.25\exp(-5(a^2 + b^2))\right\}, \tag{6.12}$$

which is the maximum of three Gaussian functions in a two-dimensional input space. To extend this function to n dimensions, the function inputs a and b are spread over half the input space respectively, written as

$$a = \frac{1}{\lfloor n/2 \rfloor} \sum_{i=1}^{\lfloor n/2 \rfloor} x_i, \quad b = \frac{1}{\lceil n/2 \rceil} \sum_{i=\lfloor n/2 \rfloor+1}^{n} x_i. \tag{6.13}$$

Henceforth, f_2 is also termed *cross* function, while the former f_1 is called the *sine* function. Both are illustrated in Figure 6.9 for $n = 2$.

While XCSF quickly learns accurate approximations in the two-dimensional case with default settings, the learning task becomes increasingly difficult in higher dimensions. The previous section discussed worst case scenarios, where every dimension has the same relevance for the function output resulting in an

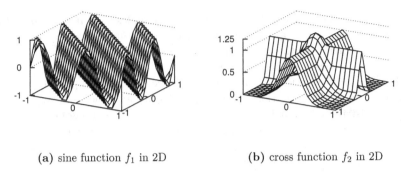

(a) sine function f_1 in 2D   (b) cross function f_2 in 2D

Figure 6.9: Two non-linear benchmark functions for XCSF. Here $n = 2$ for illustrative purpose.

exponential scaling with respect to the dimension. However, when XCSF is equipped with linear predictors and a general ellipsoidal kernel structure, the actual problem complexity is less than exponential, although all n dimensions contribute to the function value.

Function f_1 is non-linear in the direction $(1, 1, \ldots, 1)$, but all orthogonal directions remain constant. Thus, f_1 actually represents a one-dimensional, that is linear, problem in terms of population size for XCSF with a rotating RF structure and linear predictors. Optimal RFs are shrunk in the non-linear dimension and extended (beyond the input space or even to infinity) along orthogonal dimensions. Analogously, f_2 defines two non-linear parts in half the input space for a and b respectively. Consequently, the cross function poses a quadratic problem in terms of population size for XCSF. The critical requirement, however, is that the GA has actually found such an *optimal* structure and this is where the learning time comes into play.

Experiment 1: The Sine Function in 6D

As a first experiment the sinusoidal function f_1 with $n = 6$ is taken as a benchmark for XCSF. The desired target error is $\varepsilon_0 = 0.01$ while the function values range from -1 to $+1$. The maximum population size was restricted to 6400 RFs over a learning time of 100 k iterations[6]. Each RF stores up to 600 samples in

---

[6] Mutation probability was set relative to the dimensionality n to $\mu = 1/l$, where $l = n(n + 3)/2$ is the number of alleles of a rotating ellipsoidal RF. Other parameters were set to default values used before: Initial RF width is taken uniformly random from $[0.1, 1)$. $\beta = 0.1$, $\nu = 5$, $\chi = 1$, $\tau = 0.4$, $\theta_{GA} = 50$, $\theta_{del} = 20$, $\theta_{sub} = 20$. GA subsumption was applied. Uniform crossover was applied.

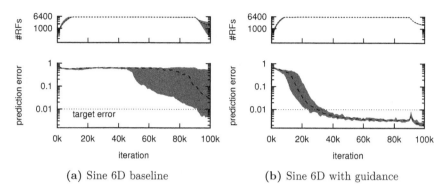

(a) Sine 6D baseline   (b) Sine 6D with guidance

Figure 6.10: XCSF's prediction error and population size in twenty independent runs on f_1 with $n = 6$. The shaded area represents worst and best case and the line depicts median performance, respectively. (a) Baseline XCSF most likely fails to reach the target error in the given learning time. (b) The guided mutation operator drastically speeds up the learning process.

its personal history, The number of stored samples is chosen rather large here, but often the GA is activated before that number is actually reached.

Twenty independent runs were executed and worst, median, and best performance are shown in subsequent graphs. Figure 6.10a shows that regular XCSF eventually reaches the desired target error in two out of twenty runs, but would require more training in the other 18 cases. When the guided mutation operator is executed with 50% chance, the target error is reached after roughly 30 k iterations (Figure 6.10b). Thus, guided mutation results in a dramatic speed-up in terms of required iterations.

6.2.3 Experiment 2: A 10D Sine Wave

When the dimensionality is increased to $n = 10$, the evolutionary search space becomes 65-dimensional for a typical, rotating RF structure. Without close-to-optimal initialization the chances of baseline XCSF to actually find the proper RF structure become extremely low and the prediction error stays at a very high level throughout all of the twenty runs, even with an increased learning time of 200 k iterations (Figure 6.11a).

The error does not drop at all because the GA cannot find a fitness gradient – without fitness pressure the *set* pressure takes over and over-sized RFs reproduce more than smaller ones. Eventually, a good choice of initial RF parameters could solve this problem: When initial RFs require few mutations to become better in terms of fitness, the GA will most likely find that niche. On the other

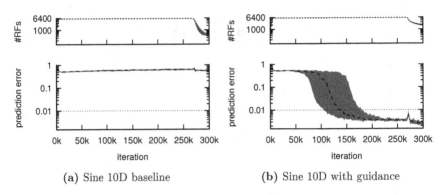

(a) Sine 10D baseline   (b) Sine 10D with guidance

Figure 6.11: The function f_1 with $n = 10$ poses a challenging problem. (a) Baseline XCSF is unable to find a suitable solution. (b) The new mutation operator guides the GA towards a suitable clustering of the high-dimensional input space.

hand, when an RF needs to be rotated in a particular plane, shrunk in $n - 1$ dimensions, and extended the remaining dimension, the chances to randomly hit such a structure become critically low.

The guided mutation operator can help to find such a small niche even in a high-dimensional evolutionary search space. The chance to activate guided mutation is set to 50% as before, but each RF can store up to 1000 samples to compensate for the increased dimensionality. Again, XCSF with guided mutation is able to reliably reach the target error throughout all twenty runs in Figure 6.11b after roughly 140 k learning steps.

Experiment 3: The Crossed Ridge Function in 10D

On the one hand, the sine function required a rather special clustering, that is, the GA has to find a small niche in the evolutionary search space. On the other hand, the same structure (thin along the sine wave, spread otherwise) can be applied throughout the full input space and, thus, the structure can be propagated effectively by means of crossover, once found. The function f_2 instead requires several different structures and, on top of that, poses a quadratic problem in terms of required population size, while the population size scales linearly in the dimension n on the sine function. Consequently, the GA has better chances to find locally optimal solutions, but every location in input space requires slightly different structures and therefore the overall balance between coverage, generality, accuracy, and low overlap is difficult to achieve.

The population limit is raised to 10 k RFs and the learning time is increased to 600 k to compensate for the difficult task. Indeed, Figure 6.12a suggests that

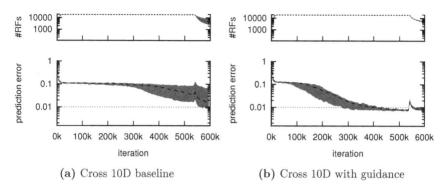

(a) Cross 10D baseline (b) Cross 10D with guidance

Figure 6.12: The function f_2 with $n = 10$ requires extensive training. (a) Baseline XCSF slowly approaches the desired accuracy. (b) Guided XCSF is able to reliably reach the target error in less than 450 k iterations.

XCSF would probably evolve an accurate solution at some time, but require even more learning time to do so. On the contrary, guided mutation helps to reach an accurate representation after about 420 k iterations (Figure 6.12b).

6.2.4 What is the Optimal Guidance Probability?

Since the developed operator is a heuristic, it cannot achieve an optimal clustering on its own – a 100% guidance probability will only work on very simple problems where the GA finds optimal structures with few regular mutations and crossover is sufficient to explore good axis lengths. Up to now, the guided mutation was applied with a 50% chance, which works well, but seems arbitrary.

To further investigate the effect of guidance the same functions are tested with different values for guidance probability ranging from 0% (baseline) to 100%. A value of 100% does not mean that guided mutation is applied exclusively, because the computed covariance matrix may be ill-conditioned and regular mutation is used as a fallback. To compare different guidance levels, the *iteration* where XCSF reaches the target error for the first time is recorded. If the target error is not reached in one run, a failure is recorded as 100% learning time. Twenty independent runs are executed for every function and probability setup and worst, median, and best performance is recorded. Figure 6.13 shows the results for all three benchmark functions in one graph, where the required learning time (until the target error is reached) is depicted as a percentage.

The original XCSF without guidance is outperformed by guided XCSF with reasonable guidance probability. While the optimum is slightly different for the three functions, a value of 50% seems reasonable without further domain

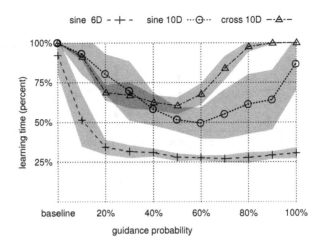

Figure 6.13: The optimal guidance probability depends on the problem at hand. Aggressive guidance works well for lower dimensional problems, but complex problems require both guided *and* regular mutation for a minimal training time. Here, 100% learning time indicates that the target error was *not* reached within the given learning time.

specific knowledge. Lower dimensional problems can be solved well even with a very high guidance probability, where the limited exploration capabilities of crossover suffice for fine-tuning of the axis lengths. However, using guidance almost exclusively on complex problems (sine 10D, cross 10D) can deteriorate the learning process.

6.2.5 Summary and Conclusion

Problem dimensionality and the RF structure used in XCSF define the dimensionality of the evolutionary search space and, thus, account for large parts of the problem difficulty. When a sophisticated RF structure, such as rotating ellipsoidal kernels, is applied to a high-dimensional problem, regular XCSF may require extensive training or may even fail completely. This is because the GA has a multitude of mutation alternatives, where few of them significantly change the fitness and mutations in multiple alleles may even cancel each other. The fitness and mutations in multiple alleles may even cancel each other.

The developed *guided mutation* operator is a heuristic that aims to both *generalize* where possible and *specialize* where necessary. Therefore a bounded history of samples is stored for every RF which, upon mutation, is used to assess the predictive quality of the RF in different directions. The computation of a weighted covariance matrix yields a shape similar to the parent RF, but extended in directions of accuracy and shrunk where large prediction errors are made. This allows to solve even complex, ten-dimensional problems, where original XCSF was unable to find a fitness gradient and the evolutionary search was lost.

Guided mutation is most useful, when the evolutionary search space is high-dimensional and the problem at hand benefits from a sophisticated clustering as opposed to spherical RFs as used in RBFN. In other words, the longer the evolutionary search for an optimal solution takes, the more local search operators can help to speed up this process.

Part III

Control Applications in Robotics

7 Basics of Kinematic Robot Control

The third part of this work is devoted to simulated robot devices, learning models of their kinematic structure, and using these models for simple directional control tasks such as reaching for objects. Learning is realized by algorithms that mimic brain function at least to some degree. Therefore the framework developed herein *could* explain how the brain learns motor control. Of course, there is no proof because a concrete implementation in one or the other programming language is far from being comparable to brain imaging results that merely highlight activity in certain regions for certain tasks. Nonetheless, this work tries to make a connection from neuron level (neurobiology) to the functional, cognitive level (psychology).

Chapter 7, the first chapter of this part, introduces robot devices and the background knowledge required for simple *kinematic* control. Kinematics describe the motion of bodies, that is, position, velocity, and acceleration. *Dynamics*, on the other hand, are concerned with forces acting on the bodies, e.g. friction, momentum of inertia, and external forces such as gravity. Dynamics are not considered in this work for several reasons: First, current robots often come with sophisticated low-level dynamic controllers and swiftly realize the kinematic commands of a user. Second, dynamics is where animals and robots are very different because muscles are not comparable to the servo motors of typical robots. Third, dynamics are better expressed as a set of differential equations than as a function.

Robots are usually an assembly of limbs, joints, and some sort of actuators – servo motors typically, but pneumatic elements are possible as well. Physical bodies are connected by either rotational (or translational) joints. The physical construction of limbs and joints is a so called kinematic chain as for example illustrated in Figure 7.1. The number of joints defines the Degrees of Freedom (DoF) of a kinematic chain. For example, the kinematic chain in Figure 7.1 has two DoF and the human arm has seven DoF.

A very basic property of a kinematic chain is that its state is uniquely defined by the state of the joints, where *state* means position and its derivatives: the first derivative of position is velocity, the second one is acceleration. Simply put, location and orientation of the end effector or any other part of the kinematic

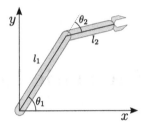

Figure 7.1: A kinematic chain composed of two limbs (and an imaginary base) with two rotational joints. The (x, y) position of the gripper is defined by joint angles θ_1 and θ_2.

chain can be computed from the joint angles. The so called forward kinematics are covered in Section 7.1.

A certain task does not always require all joints, or more specifically, can sometimes be solved with multiple joint configurations. While this may seem like a problem, it is actually more of an opportunity. The so called kinematic *redundancy* allows to reach the same task location with different joint configurations – which helps to fulfill secondary constraints such as obstacle avoidance, reducing energy consumption, or maintenance of comfortable joint states. On the contrary, it is sometimes impossible to produce a particular task space trajectory due to so called *singularities* at certain joint configurations. Both kinematic redundancy and singularities are the topic of Section 7.2.

In order to realize a given task, a controller need not know about the forward kinematics but instead requires the *inverse* formulation: What joint state is required for a desired task state? Section 7.3 discusses the inverse kinematics and how to smartly invert the forward mapping while not only avoiding singularities but also incorporating other constraints.

Finally, given a full task and a robot device, the controller must *plan* a trajectory from current to target location. However, sophisticated planning is not the topic of this work and a very simple directional scheme is employed in Section 7.4 by simply moving in a straight line to the target. To summarize briefly, the present chapter covers the basics of kinematic robot *control*, but *learning* a kinematic model is discussed in the follow-up chapter.

7.1 Task Space and Forward Kinematics

Usually the task of a robot cannot be described in terms of joint angles. Instead, a task will often be described by Cartesian coordinates of the end effector, eventually together with an orientation. However, any other task space formulation is possible, as long as a certain joint configuration uniquely defines the task space configuration. This section establishes the notation for control space (denoted by capital theta, Θ) comprised of all possible joint configurations, task space (denoted by capital xi, Ξ) consisting of corresponding task states, and the relation between them.

The control space $\Theta \subset \mathbb{R}^n$ has dimension equal to the robots DoF[1]. A given joint configuration $\boldsymbol{\theta} \in \Theta$ uniquely defines the

- Cartesian position of joints,

- Cartesian position of any other fixed part of the kinematic chain, e.g. the end effector, and

- Orientation of kinematic elements (limbs).

Thus, the task space $\Xi \subset \mathbb{R}^m$ could be the pose (position and orientation) of the end effector, which allows a suitable formulation of reaching tasks. However, other formulations may be used as well as long as task space location is uniquely defined by control space location via the so called forward kinematics equation

$$\boldsymbol{\xi} = f(\boldsymbol{\theta}), \qquad (7.1)$$

which is also called *zeroth order* forward kinematics. The function f is a typically non-linear mapping that depends on the location of rotation axes, the location of limbs relative to those axes, and the actual rotation angles often denoted as joint angles. The forward kinematics of the planar 2 DoF robot displayed in Figure 7.1, with respect to the Cartesian location of the gripper, are given as

$$
\begin{aligned}
x &= l_1 \cos(\theta_1) + l_2 \cos(\theta_1 + \theta_2) \\
y &= l_1 \sin(\theta_1) + l_2 \sin(\theta_1 + \theta_2).
\end{aligned} \qquad (7.2)
$$

The next thing to talk about is velocity, which is the derivative of position with respect to time (the difference in position per time unit). The first derivative of Equation (7.1) with respect to time yields the *first order* forward kinematics

$$\dot{\boldsymbol{\xi}} = \boldsymbol{J}_f(\boldsymbol{\theta})\dot{\boldsymbol{\theta}}, \qquad (7.3)$$

that is, how joint velocities $\dot{\boldsymbol{\theta}}$ are mapped to task space velocities $\dot{\boldsymbol{\xi}}$. The so called Jacobian matrix $\boldsymbol{J}$ depends on the actual kinematic relation f and on the current joint configuration $\boldsymbol{\theta}$. This is always the case and the symbol $\boldsymbol{J}$ is a short form of the more explicit $\boldsymbol{J}_f(\boldsymbol{\theta})$. Given a particular joint configuration, the Jacobian $\boldsymbol{J}$ defines a linear relation from joint velocities to task velocities. Figure 7.2 illustrates the effect of rotational joint velocity. It is also possible to describe accelerations, that is, second order forward kinematics. However, those are not used throughout this thesis.

---

[1] Exceptions are robots that have multiple actuators for a single joint, e.g. two pneumatic artificial muscles.

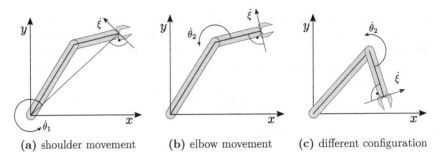

(a) shoulder movement   **(b)** elbow movement   **(c)** different configuration

Figure 7.2: The forward velocity kinematics map joint angle velocity $\dot{\boldsymbol{\theta}}$ to task space velocity $\dot{\boldsymbol{\xi}}$, which is tangential to the rotation circle. (a) The shoulder joint has the strongest impact because of its distance to the gripper. (b) The elbow rotation produces task space velocity in a different direction. (c) The actual joint configuration $\boldsymbol{\theta}$ defines possible directions.

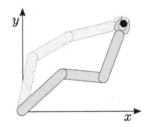

Figure 7.3: A certain end effector location can be realized with different, redundant joint configurations. When the robot device has more DoF than the task space dimension, there is usually an infinite number of distinct configurations corresponding to one task space location.

7.2 Redundancy and Singularities

When the task space dimensionality $|\Xi|$ is less than the control space dimensions $|\Theta|$, a kinematic chain is said to be *redundant* with respect to its task as illustrated in Figure 7.3. For example, reaching for a cup of coffee requires six DoF in the human arm (3D position and 3D orientation), but the human arm has seven DoF. Typically, one can move the elbow freely without moving the cup – this redundancy allows to take other constraints into account, e.g. comfort.

Mathematically, redundancy can be best formulated with the Jacobian matrix from Equation (7.3). If the Jacobian's rank is greater than the task space dimensionality, there is redundancy that can be used otherwise. Then, there exists the so called nullspace, which is the space of movements that do not

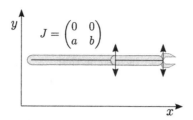

Figure 7.4: It is impossible to create a non-zero x task space velocity in this singular configuration, because both joints produce velocity exclusively in y direction.

interfere with the current task space location. Secondary constraints can be projected into the nullspace and consequently do not affect the current primary task.

7.2.1 Singularities

Contrary to redundancy, there are some rare joint configurations where the Jacobian cannot generate a certain task space direction. An example is shown in Figure 7.4 where the 2 DoF arm can not move its end effector in the x direction because the velocities generated by all joints are linearly *dependent*. Put differently, the Jacobian matrix is rank deficient, that is, its rank is less than the dimension of the task space. The velocity direction vectors in Figure 7.2 are linearly *independent* and consequently task space movement in any direction is possible (when joint angle boundaries are ignored).

While the example above shows an exactly singular configuration, the problem is also present close to such configurations, where almost infinite joint velocities are required to produce a decent task space movement. There are many different strategies to prevent singular configurations. For example, available redundancy can be used to stay away from singular configurations by using a singularity-distance metric. However, when aiming for an object that is out-of-reach, singular configurations are almost inevitable. Thus, control strategies should not only *avoid* singularities [4], but should also be able to *cope with* singularities.

To sum up, the rank of the Jacobian matrix specifies if the system is in a redundant state (rank r greater than the task space dimension m), a singular state ($r < m$), or if there exists a single trajectory for the task at hand ($r = m$) [22, 5, 79]. In all these case, an effective controller must be able to find a path towards the current task space target. A particularly elegant solution is discussed in the following section.

7.3 Smooth Inverse Kinematics and the Nullspace

Section 7.1 described the forward kinematics that map from control space onto task space. For this section, let us assume that the forward model of an n DoF robot device acting on an m-dimensional task space is known. Put differently, the Jacobian matrix is known for any joint configuration.

In order to control the robot such that a certain target is reached, an inverse formulation of the forward kinematics is required: A desired task space movement $\dot{\xi}^*$ is given and corresponding control space velocities (the required control commands) are unknown. Thus, Equation (7.3) has to be solved for the control space velocity

$$\dot{\theta}^* = J^{-1}\dot{\xi}^*, \tag{7.4}$$

but it is not clear what the inverse Jacobian J^{-1} should be, because that system of equations may be over- or under-determined (singular or redundant), and rarely has a unique solution. This is the so called *problem of* inverse kinematics.

The *Moore-Penrose* matrix or Pseudoinverse [7] provides a single solution to the inversion of a system of linear equations. If the the system is under-determined the Pseudoinverse provides a solution with minimum norm out of the multitude of possible solutions. Even if the system is over-determined (singular), the Pseudoinverse matrix yields a solution that minimizes the squared error, that is, the closest possible solution to the inversion. This is not only used for inverse kinematics [58], but has many other applications, e.g. the Ordinary Least Squares (OLS) method described earlier in Section 2.3.1.

However, minimizing the squared error in near-singular configurations is actually not desired: The resulting velocities from the regular Pseudoinverse become extremely large, which is not feasible for a real robot. Furthermore, there is a gap from near-singular $\lim_{x \to 0} 1/x = \infty$ to the exact singular position $1/0$ which is treated as zero in the Pseudoinverse matrix. The controller should act in a smooth way while moving through the control space, instead of jumping from almost infinite velocity to zero velocity. Moreover, actual implementations in hardware have a limited numerical precision and, thus, become unstable towards singularities.

For all those reasons it is particularly convenient to know *how singular* the current Jacobian matrix is and to act accordingly, which is not possible with a simple Pseudoinverse. Singular Value Decomposition (SVD) [33, 66] offers an all-in-one solution to inverse kinematics, because a) it provides the level of singularity, b) it can be used to solve the inverse problem in a smooth way, and c) the SVD additionally provides the nullspace, which allows to incorporate secondary constraints to resolve available redundancy.

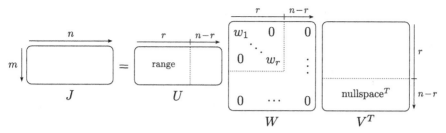

Figure 7.5: Given an $m \times n$ matrix J, the SVD computes three matrices of different dimensions. U has the same size as the given matrix, while W and V are $n \times n$ matrices. Up to $r \le m, n$ non-zero singular values are present on the diagonal of W. Thus, J is of rank r. Correspondingly, the first r columns of U span the range of J and the nullspace is of dimension $n - r$, spanned by the last columns of V (or rows of the transposed matrix).

7.3.1 Singular Value Decomposition

Let J be the $m \times n$ Jacobian matrix for the current robot configuration with an n-dimensional control space and m-dimensional task space. Typically, m is less or equal to n in robot control scenarios[2]. Furthermore, let

$$J = UWV^{T} \qquad (7.5)$$

the SVD of the Jacobian matrix where U is a $m \times n$ column-orthogonal matrix, W is a $n \times n$ diagonal matrix of singular values, and V^{T} is the transpose of an $n \times n$ orthogonal matrix [66]. Usually, the singular values w_i are sorted in descending order such that the first $k \le m, n$ values are non-zero. A singular configuration is present when one of the largest m singular values is zero, that is $k < m, n$. Then the range of J

$$\{\dot{\xi} \in \dot{\Xi} \mid \dot{\xi} = J\dot{\theta}, \, \dot{\theta} \in \dot{\Theta}\}, \qquad (7.6)$$

that is, the space of all producible task space velocities, is smaller than the actual velocity task space and, thus, certain directions can not be generated. The transposed matrix V contains range and nullspace as depicted in Figure 7.5.

The SVD is of great use here. First, it is straight forward to detect singular or near-singular configurations by looking at the singular value at index m. Second,

---

[2]Many formulations of SVD require the opposite, that is $m \ge n$, but this is actually not required. The most influential work for implementations of SVD was written by Golub and Reinsch [34] and starts with the sentence "Let A be a real $m \times n$ matrix with $m \ge n$.". Concrete implementations for $m < n$ are available in [28, 66].

the nullspace is readily available for projection of secondary constraints, which is explained later. Third, it is easy to compute an inverse of J, be it the standard Pseudoinverse or an adaptive inverse depending on the level of singularity.

7.3.2 Pseudoinverse and Damped Least Squares

Given the decomposition $J = UWV^T$, the Pseudoinverse $J^\dagger$ is easily computed as

$$J^\dagger = VW^\dagger U^T , \qquad (7.7)$$

where the Pseudoinverse of W is is formed by replacing non-zero diagonal elements by its reciprocal. This results in the *least squares* solution to the inversion of J. Values $w_i < \epsilon_{mach}$ are treated as zero due to numerical precision, where ϵ_{mach} is an upper bound for rounding errors, e.g. 2^{-52} for 64 bit floating point numbers. The resulting Pseudoinverse could then be used in Equation (7.4) as

$$\dot{\theta}^* = J^\dagger \dot{\xi}^*$$
$$= \sum_{i=1}^{r} \frac{1}{w_i} v_i u_i^T \dot{\xi}^* , \qquad (7.8)$$

where $\dot{\xi}^*$ is the desired task space velocity and r is the rank of J, that is, the number of non-zero singular values w_i.

However, near-singular configurations will produce near-infinite velocities to compensate for the desired, near-impossible movement direction. An exemplary unit of meter per second and the inverse value $2^{52} \approx 4.5 \cdot 10^{15}$ m/s seems inappropriate for any robot. In the following, different, increasingly sophisticated solutions for near-singular configurations are discussed.

The simplest solution is to use a comparably large cut-off value λ instead of the machine epsilon ϵ_{mach}. However, this still results in a sudden jump in control behavior, once a singular value goes below the cut-off value. An alternative solution was proposed by [63, 93] independently as follows. Instead of using the plain reciprocal $1/w$ in Equation (7.8), a *damped* inverse

$$\frac{w}{w^2 + \lambda^2} \qquad (7.9)$$

is applied. For values $w \gg \lambda$ the dampening becomes negligible. *Fixed dampening* converges to zero smoothly for $w \to 0$, as shown in Figure 7.6.

This method yields a valid solution for any configuration and, most importantly, the control behavior is smooth for transitions from near-singular to singular configurations. The downside is that such a *damped* Pseudoinverse is always inaccurate, even for non-singular configurations. Ideally, the dampening

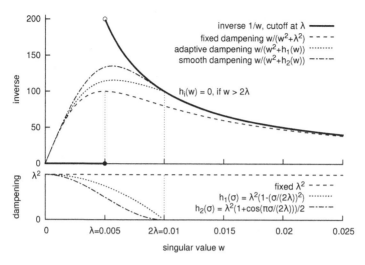

Figure 7.6: Four strategies to handle singularities are illustrated. The default for Pseudoinverse solutions is to cut-off small values (thick line). A fixed dampening yields a smooth transition to zero at the cost of accuracy. Adaptive dampening resolves the inaccuracy for non-singular solutions and the step to a smooth, differentiable solution is small.

would kick-in only when a singularity is approached. This can be realized by an *adaptive* dampening [20] via

$$h_1(w) = \begin{cases} \lambda^2 \left(1 - \left(\frac{w}{2\lambda}\right)^2\right) & \text{if } w < 2\lambda, \\ 0 & \text{otherwise.} \end{cases} \tag{7.10}$$

This results in a smooth transition from near-singular configurations at 2λ towards zero and Equation (7.9) then becomes

$$\frac{w}{w^2 + h_1(w)} . \tag{7.11}$$

Singular values greater than 2λ are not affected and the standard reciprocal $w/(w^2 + 0) = 1/w$ is applied as is.

The disadvantage of the *fixed dampening* is resolved, but a small issue remains. While there is no gap in the velocity produced by this formula, the transition at 2λ is not differentiable. In other words, there is a gap in the first order derivative, which is acceleration. Typically this is not a problem and is also not discussed in [20]. However, it is easy to come up with a smooth and differentiable solution by use of a sigmoid transition.

Throughout the remaining experiments with inverse kinematics, the following *smooth* dampening is applied:

$$h_2(w) = \begin{cases} \lambda^2 \left(0.5 + 0.5\cos(\frac{\pi w}{2\lambda})\right) & \text{if } w < 2\lambda\,, \\ 0 & \text{otherwise}\,, \end{cases} \tag{7.12}$$

which is also depicted in Figure 7.6.

To put everything together, the solution to the inverse kinematics problem in Equation (7.4) is computed as

$$\dot{\boldsymbol{\theta}}^* = \boldsymbol{J}^\dagger \dot{\boldsymbol{\xi}}^*$$
$$= \sum_{i=1}^r \frac{w_i}{w_i^2 + h_2(w_i)} \boldsymbol{v}_i \boldsymbol{u}_i^T \dot{\boldsymbol{\xi}}^*\,, \tag{7.13}$$

which is equal to the least squares Pseudoinverse solution for non-singular configurations (all $w_i \geq 2\lambda$). Smooth dampening becomes increasingly active towards singular configurations ($w_i < 2\lambda$) to assure smooth, low-velocity transitions to singular states. The adaptive and smooth dampening allows to handle non-singular, near-singular, and singular solutions with a single formula[3]. The value $\lambda = 0.005$, as displayed in Figure 7.6, is used throughout all experiments in this work.

7.3.3 Redundancy and the Jacobian's Nullspace

Now the robot can be controlled to move in certain task space directions. The current solutions favor minimum joint movement in the least squares sense, but available redundancy is not yet used. Humans, however, can move their elbow without changing the pose of their hand by taking additional constraints into account. Those control space movements that *do not* affect the task space state can be best understood by the nullspace of the Jacobian, which is defined as

$$\{\dot{\boldsymbol{\theta}} \in \dot{\Theta} \mid \boldsymbol{J}\dot{\boldsymbol{\theta}} = 0,\ \dot{\boldsymbol{\theta}} \in \Theta\}\,. \tag{7.14}$$

Secondary constraints such as a comfortable posture, avoiding joint limits, or minimal motor effort can be projected onto the nullspace and do not affect the actual location in task space. Eventually, a constraint can be fully achieved in parallel to the primary movement or it conflicts with the primary task and projection onto the nullspace completely nullifies the constraint. Since SVD

---

[3]Theoretically, a robot device *could* reach an exactly singular configuration ($w_i = 0$) and the controller *could* request exactly the impossible direction, which would result in zero movement – the robot got stuck. However, due to sensor & motor noise, limited numerical precision, and due to the low chance that this actually happens it can be ignored.

gives direct access to the nullspace (see Figure 7.5), this is another point why SVD excels for inverse kinematics[4]. The simplest constraint possible is $\dot{q}_0 = 0$, which basically says: "Don't move at all". This does not modify the Pseudoinverse solution, which minimizes joint velocity by default. Alternatively, the controller may strive for a comfortable or default posture θ_{default}, which can be realized by setting

$$\dot{q}_0 = \frac{v}{\|\theta_{\text{default}} - \theta\|} \left(\theta_{\text{default}} - \theta\right), \qquad (7.15)$$

where v is the desired velocity. This may also help to avoid singularities, when the preferred posture is far from being singular. Another approach is to directly avoid singularities via a measure of singularity, which is then converted into a favorable direction [79]. However, with the smooth, damped inverse to cope with singularities, the redundancy can be used otherwise.

The actual projection to the nullspace works as follows. Let $J = UWV^T$ be the SVD of $J \in \mathbb{R}^{m \times n}$ with rank r. Consequently, the last $n - r$ columns of V represent the nullspace of the Jacobian. Therefore, let $O = (v_{r+1} \ldots v_n)$ be the matrix comprised of the last $n - r$ columns of V. A given constraint $\dot{q}_0 \in \dot{\Theta}$ is then projected onto the nullspace via

$$OO^T \dot{q}_0 = \sum_{i=r+1}^{n} v_i v_i^T \dot{q}_0. \qquad (7.16)$$

Combined with Equation (7.13) this gives the full formula for inverse kinematics including adaptive, smooth dampening for singularities and redundancy exploitation to incorporate secondary constraints. The Jacobian J for the current control space configuration $\theta \in \Theta$ is first decomposed via SVD into $J = UWV^T$, which yields the rank r, non-zero singular values $w_1, \ldots, w_r$, and the nullspace basis $O = (v_{r+1} \ldots v_n)$. Given a desired task space velocity $\dot{\xi}^*$ and an additional control space velocity constraint $\dot{q}_0 \in \dot{\Theta}$, the required control command is computed as

$$
\begin{aligned}
\dot{\theta}^* &= J^\dagger \dot{\xi}^* + OO^T \dot{q}_0 \\
&= \sum_{i=1}^{r} \frac{w_i}{w_i^2 + h_2(w_i)} v_i u_i^T \dot{\xi}^* + \sum_{i=r+1}^{n} v_i v_i^T \dot{q}_0, \qquad (7.17)
\end{aligned}
$$

---

[4]It is difficult to find a open source implementation of a full SVD (as opposed to thin SVD). To the best of my knowledge, only the original Fortran implementation of LAPACK [2] includes the full variant, while all newer derivatives only contain the thin versions. Eventually, even $m \geq n$ is required, which further complicates the work. A notable exception is OpenCV [10], which provides the full SVD in C, C++, and Python as a mapping to the Fortran code of LAPACK. Eventually, Octave [27] works as well by setting the "driver" accordingly, which is nothing more than the appropriate mapping to Fortran code of LAPACK.

where h_2 is the dampening function from Equation (7.12), the first summand defines the control commands to realize the desired task space velocity, and the second summand refers to nullspace movement to fulfill the secondary constraint while not disturbing the primary task space movement. The following section incorporates the above formula into a simple directional control scheme.

7.4 A Simple Directional Control Loop

Closed loop control consists of three steps: a) The current position is measured, b) a planning algorithm proposes a trajectory to a given target, and c) the corresponding control commands are sent to the robot device. Sophisticated planning, collision detection, and obstacle avoidance is beyond the scope of this thesis. Instead a simplified directional planning is applied that proposes a straight line in task space from current to target position.

Suppose that a robot device shall move towards a task space target $\boldsymbol{\xi}^*$, while its current task space location is $\boldsymbol{\xi}$. Then

$$\Delta\boldsymbol{\xi}^* = \boldsymbol{\xi}^* - \boldsymbol{\xi} \tag{7.18}$$

is the direct movement which assumes that the target can be reached within a single step. Before the resulting $\Delta\boldsymbol{\xi}^*$ is plugged into Equation (7.17), it must be converted to an appropriate velocity $\dot{\boldsymbol{\xi}}^*$ – that is, displacement per time unit in contrast to absolute displacement.

This can can be realized by means of a *velocity profile*, which typically specifies a linearly increasing velocity over time – capped at a certain maximum – and a linear deceleration towards the target position as shown in Figure 7.7a. Resulting movements are smoothly converging towards their final position. For the present work, however, control is not based on time but instead based on the distance to the target. The traveled distance is computed as the integral (over time) of the velocity profile. The reversed integration from the point in time, where the target is reached, towards the beginning of the movement yields the distance to the target (Figure 7.7b). In turn, the target distance can be plotted against the velocity (Figure 7.7c). Such a profile can be used to compute suitable velocities depending on the remaining distance to the current target. Alternatively, PID control can be employed as well.

However, such a velocity profile is only required when dealing with real physics. When a purely kinematic simulation is used, the result from Equation (7.18) can be capped to a maximum velocity and applied as is otherwise, because acceleration or deceleration are not taken into account. The complete control loop employed here looks as follows.

1. Retrieve current task space state of the robot device and compute the desired task space direction $\Delta\boldsymbol{\xi}^*$ towards the target via Equation (7.18).

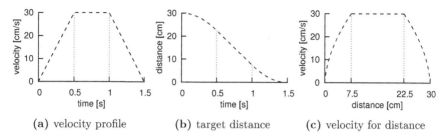

(a) velocity profile   (b) target distance   (c) velocity for distance

Figure 7.7: An appropriate velocity profile grants smooth movements. Here, the acceleration takes 0.5 s up to a maximum of 30 cm/s. (a) Typically the velocity is increased linearly over time and then capped to a maximum. Deceleration towards the target is linear as well. (b) The corresponding position changes smoothly. Here the distance to a virtual target is depicted. (c) If control is based on the target distance (as opposed to time), the acceleration and deceleration profiles become non-linear.

2. If a purely kinematic device is present, the desired velocity is

$$\dot{\xi}^* = \max\left(\Delta\xi^*, \dot{\xi}_{max}\right) \, ,$$

otherwise a suitable velocity is computed by means of a velocity profile or PID controller. The primary task is to realize the resulting task space velocity.

3. Next, retrieve the Jacobian J for the current control space configuration θ and select a secondary constraint $\dot{q}_0$.

4. After decomposition of J via SVD, Equation (7.17) computes the required control commands $\dot{\theta}^*$, which are finally executed.

This control scheme aims for a straight task space path from current location to target. Unfortunately, it is not guaranteed that a straight path is available – even without obstacles – because joint limits are not represented in the Jacobian matrix. An example is depicted in Figure 7.8, where the elbow joint can not rotate any further clockwise. Consequently, the simple directional control works only for short paths in control space, that is, reasonable joint rotations well below 180°.

A more sophisticated planning is beyond the scope of this thesis. Instead, the focus lies on *learning a kinematic model*, that is, learning the Jacobian matrix for any desired robot configuration.

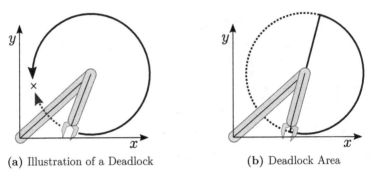

(a) Illustration of a Deadlock   (b) Deadlock Area

Figure 7.8: The task is to move the gripper to the target (cross), but the elbow joint can not rotate any further clockwise. (a) The Jacobian suggests the shortest path in control space (dashed), but does not account for joint limits. Instead, the valid path to the target is the long route in the opposite direction and initially increases the distance to the target. (b) Chances for deadlocks are rather large, when task space targets are generated randomly.

8 Learning Directional Control of an Anthropomorphic Arm

The components required for a successful robot control framework are ready to be put together. Velocity control, as described in the previous chapter, is based on the Jacobian matrix and the first section of this chapter illustrates how to apply Function Approximation (FA) algorithms to *learn* the Jacobian, which essentially defines the forward kinematics. In turn, Section 8.2 combines *learning* and *control* into a complete framework.

Next, Section 8.3 defines the structure of an anthropomorphic seven Degrees of Freedom (DoF) arm, the corresponding task space, and how tasks are generated for training and test stages. The subsequent Section 8.4 discusses methods to assess the performance of a learned model.

The experimental Section 8.5 starts with a baseline experiment, where a single linear model approximates the complete forward kinematics. This serves three purposes: First, it shows that a linear model is not sufficient to approximate the non-linear kinematics, but it also reveals that the learning task is not too difficult. Finally, a single linear model represents the worst case performance that any non-linear FA algorithm should outperform.

Next, the three FA algorithms identified in Chapter 2 are applied to learn the velocity kinematics of the anthropomorphic arm. The first algorithm, Radial Basis Function Network (RBFN), is the basic neuron-based approach to approximation with a fixed set of Receptive Fields (RFs). Thus, only the local models are *learned* here, while number, position, and shape are defined by the user. The second candidate, XCSF, is based on a Genetic Algorithm (GA) to find an optimal structuring of the population of RFs. The third is Locally Weighted Projection Regression (LWPR), which creates RFs on demand and further adapts their shape to reduce the prediction error. The optimization is based on a heuristic gradient descent. Experiments not only evaluate the accuracy of kinematic control, but briefly test the knowledge of redundant alternatives as well.

8.1 Learning Velocity Kinematics

The essential element that defines the forward kinematics in Equation (7.3) is the Jacobian matrix $\boldsymbol{J}$. In the context of a certain joint configuration, this matrix defines a linear relation $\dot{\boldsymbol{\xi}} = \boldsymbol{J}\dot{\boldsymbol{\theta}}$ between control space velocity $\dot{\boldsymbol{\theta}}$ and task space velocity $\dot{\boldsymbol{\xi}}$. For an n-DoF arm and an m-dimensional task space, the resulting system of equations looks as follows.

$$\dot{\xi}_1 = J_{11}\dot{\theta}_1 + \ldots + J_{1n}\dot{\theta}_n$$

$$\vdots$$

$$\dot{\xi}_m = J_{m1}\dot{\theta}_1 + \ldots + J_{mn}\dot{\theta}_n \tag{8.1}$$

While this is indeed linear for a single joint configuration $\boldsymbol{\theta}$, the slope of the linear relation changes with the configuration. Thus, the Jacobian is best understood as a *parametrized* linear function $f_{\boldsymbol{\theta}}(\dot{\boldsymbol{\theta}}) = \boldsymbol{J}(\boldsymbol{\theta})\dot{\boldsymbol{\theta}}$.

With this in mind, the FA algorithms that learn a patchwork of locally linear models are well suited to learn the forward kinematics[1]. It remains to be shown how the *parametrized* velocity kinematics can be learned. Generally, there are three ways to accomplish this.

Either, the *zeroth* order forward kinematics from Equation (7.1) are learned and the *first* order derivative for a certain configuration is computed manually. When locally linear models are used to approximate this mapping, the first order derivative is naturally represented in the local linear models, because those approximate the gradient at a given configuration. This method is available per default in the LWPR implementation [54]. Thus, LWPR learns the zeroth order mapping $\boldsymbol{\theta} \mapsto \boldsymbol{\xi}$ and computes the derivative, that is, the Jacobian $\boldsymbol{J}$ from its local linear models surrounding the current configuration $\boldsymbol{\theta}$. Alternatively, the parametrization could be learned as an argument to the function. Consequently, the mapping $\boldsymbol{\theta} \times \dot{\boldsymbol{\theta}} \mapsto \dot{\boldsymbol{\xi}}$ is learned. This doubles the dimension of the input space unnecessarily and should not be used.

Finally, the first order kinematics can be learned directly, by separating the input for activation functions and local models. This route is taken in the implementation of RBFN and XCSF [80] used here. Figure 8.1 illustrates the required modification of the standard RBFN approach. Implementation-wise, the difference is comparably small. Equation (2.23) defined the prediction of an RBFN with m neurons as

$$h(\boldsymbol{x}) = \frac{\sum_{r=1}^{m} \phi_r(\boldsymbol{x})\lambda_r(\boldsymbol{x})}{\sum_{r=1}^{m} \phi_r(\boldsymbol{x})},$$

---

[1]The following is largely based on [83].

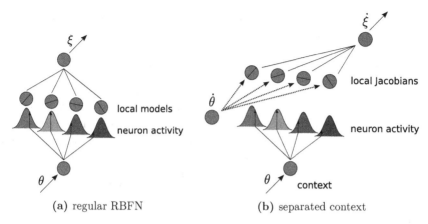

(a) regular RBFN   (b) separated context

Figure 8.1: Learning the velocity kinematics requires a subtle modification. (a) Usually, an RBFN approximates a non-linear function $\theta \mapsto \xi$. (b) When a parametrized, linear function is learned, the context θ defines which neurons are active. The parametrized velocity mapping $\dot{\theta} \overset{\theta}{\mapsto} \dot{\xi}$ is linear and is represented in the individual local models, which essentially learn the Jacobian in the vicinity of a certain configuration θ.

where ϕ_r is the activation function of the r-th neuron (kernel) and λ_r is the respective local model. Now instead the activation functions receive the current configuration θ and the local models learn the velocity mapping, written as

$$h(\boldsymbol{\theta}, \dot{\boldsymbol{\theta}}) = \frac{\sum_{r=1}^{m} \phi_r(\boldsymbol{\theta})\lambda_r(\dot{\boldsymbol{\theta}})}{\sum_{r=1}^{m} \phi_r(\boldsymbol{\theta})} . \tag{8.2}$$

The local models here are linear to represent the Jacobian in the vicinity of an kernel. As a zero joint velocity implies a zero task velocity, an intercept is not required.

Instead of querying predictions for a single data point, the velocity control requires the whole Jacobian matrix in Equation (7.17). This matrix is formed as a weighted sum

$$\boldsymbol{J} = \frac{\sum_{r=1}^{m} \phi_r(\boldsymbol{\theta})\beta_r^T}{\sum_{r=1}^{m} \phi_r(\boldsymbol{\theta})} , \tag{8.3}$$

from individual linear models β analogously to regular predictions.

The conceptual changes in XCSF are similar: activation of RFs is determined by the configuration which serves as the (zeroth order level) context and the linear prediction is trained at the (first order) velocity level. However, XCSF's predictions are typically weighted by the fitness of individual RFs, which results in a jerky control as the transition from one RF to another is not smooth.

Thus, the same activity-based weighting as in Equation (8.3) is used in XCSF to generate the Jacobian matrix for a given configuration.

8.1.1 Learning on Trajectories

Kinematic data is typically generated on trajectories, that is, the samples ($\boldsymbol{\theta}$, $\dot{\boldsymbol{\theta}}$, $\dot{\boldsymbol{\xi}}$) are close to each other: The control space configuration $\boldsymbol{\theta}$ slowly changes and the velocity often remains the same or points in the same direction for a while. This does not only prolongate the learning, but also gives misleading information to the GA in XCSF or to the statistical gradient in LWPR and a special treatment is required for accurate approximations.

An example illustrates the problem. Suppose a 7 DoF arm moves in a 3 D task space at a fixed joint velocity and a single RF gathers the underlying kinematic data. The resulting linear model (Recursive Least Squares (RLS) or similar) accurately models the data, but can only learn the current velocity. The effect of orthogonal velocities is unknown and taken to be zero (following Occam's razor). Put differently, the model is sufficiently accurate when it has seen movement in all relevant directions. However, as the robot device moves along, different RFs become active and each one only sees a single trajectory which is often rather straight. It takes a considerably large amount of training until all directions are sufficiently sampled for every individual RF.

The local RLS models are affected, but it is only a question of time until they become accurate. However, the crux is that the GA intermittently requests the prediction error to assess and compare different RFs. The prediction error on a trajectory is completely misleading, as the local model quickly adapts to small changes along a path. Thus, the fitness value is not representative and evolution favors structures that *have been* useful on the past trajectory, but probably do not accurately resemble the Jacobian for *future* movements. Similarly, the gradient descent in LWPR optimizes towards past trajectories. As the RF structure in RBFN is fixed, this algorithm is not affected.

The problem can be circumvented in different ways. First, kinematic data can be gathered *before* learning, as opposed to learning *online* during control. The order of the samples could be randomized or the data could be clustered such that all movement directions are sampled more or less uniformly. However, this does not work in an online environment and is not plausible with respect to brain functionality.

Instead, a more natural approach could be termed "learning from failures", where updates are only triggered when the prediction of the model is actually wrong. Therefore, when a sample $\boldsymbol{\theta}, \dot{\boldsymbol{\theta}}, \dot{\boldsymbol{\xi}}$ arrives, the algorithm is first asked for a prediction at the given data point. When the prediction is sufficiently accurate the sample is ignored and otherwise incorporated into the model. The term "sufficiently accurate" depends on the length of the kinematic chain and the maximum velocity of the robot device. It should be set according to the maximum error allowed for the task at hand.

8.1.2 Joint Limits

One thing that can not be represented properly in such a learned model are joint limits. Even close to a joint limit, the true Jacobian matrix represents velocities that would result in a crash. When the learning algorithm incorporates sensory measurements at a joint limit, the local models close to the limit will be skewed to represent zero velocity *towards* the border.

Even worse, as the Jacobian is a linear matrix, the *opposite* directions receive the same treatment – the matrix becomes completely singular. Subsequent control using this Jacobian might be stuck at the border. Consequently the learning algorithms does not receive updates when one joint is at one of its extremes. This preserves the correct directions for the Jacobian and the controller is able to move further towards its target. During early learning, however, control can get stuck this way. Therefore, a movement is considered as a failure after a sufficiently large number of steps.

8.2 Complete Learning and Control Framework

All parts of the framework are summarized and Figure 8.2 illustrates the six steps of the control loop.

1. At time t some kind of sensors provide the current control space state $\boldsymbol{\theta}_t$ and task space state $\boldsymbol{\xi}_t$. Noise is not taken into account, yet.

2. The controller module computes the derivatives with respect to time, that is, velocities

$$\dot{\boldsymbol{\theta}}_{t-1} = \frac{\boldsymbol{\theta}_t - \boldsymbol{\theta}_{t-1}}{\Delta t} \quad \text{and} \quad \dot{\boldsymbol{\xi}}_{t-1} = \frac{\boldsymbol{\xi}_t - \boldsymbol{\xi}_{t-1}}{\Delta t},$$

where Δt is the difference in simulation time between successive sensor measurements. It is important to clarify the causal relation: Given a configuration $\boldsymbol{\theta}_{t-1}$, the control space velocity $\dot{\boldsymbol{\theta}}_{t-1}$ results in a task space velocity $\dot{\boldsymbol{\xi}}_{t-1}$.

3. When a joint is at one of its limits or when the model's prediction error $\varepsilon = \|\dot{\boldsymbol{\xi}}_{t-1} - \boldsymbol{J}(\boldsymbol{\theta}_{t-1})\dot{\boldsymbol{\theta}}_{t-1}\|$ is sufficiently low, the kinematic model is not updated. Otherwise the learning algorithm (one of RBFN, XCSF, LWPR) receives a training sample. LWPR receives a zeroth order sample $\boldsymbol{\theta}_t, \boldsymbol{\xi}_t$,

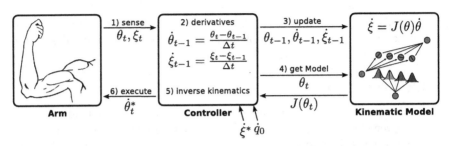

Figure 8.2: A controller module works hand in hand with an FA algorithm to learn the velocity model of a kinematic chain and to solve given tasks.

while RBFN and XCSF receive a first order sample $\theta_{t-1}, \dot{\theta}_{t-1}, \dot{\xi}_{t-1}$ as described in Section 8.1.

4. The Jacobian $J(\theta_t)$ is requested for the current configuration (Equation (8.3)).

5. The desired task space direction $\dot{\xi}^*$ and a secondary control space velocity constraint $\dot{q}_0$ are the external control signals required for inverse control (cf. Equation (7.17)). If not mentioned explicitly, the secondary constraint is a simple "avoid joint limits" velocity that always points towards a centered configuration.

6. Finally, the computed velocity command is sent to the arm and the steps are repeated.

For RBFN and XCSF steps 3 and 4 work at different points in time: the Jacobian is requested for the *current* configuration while the model update is triggered for the *previous* configuration and corresponding velocities, because velocities for the current configuration are not known, yet. On first sight, this may be a small advantage of learning the zeroth order forward kinematics as done in LWPR, but the knowledge gain is similar because velocities are missing here.

In any case there is a strong dependency between updates and model requests, as they occur at configurations close to each other. With $\Delta t \to 0$ the model would learn exactly what is required for subsequent control. Thus, to measure performance of any kind, the update step 3 should be disabled during a test phase. Then, the model is able to learn a general kinematic mapping during a training stage, but its knowledge is measured offline – without explicitly "learning the test" meanwhile.

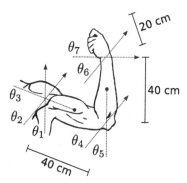

Figure 8.3: Rotation axes $\theta_1, \ldots, q_7$ are drawn as dashed lines and rotary joints θ_3, θ_5 are depicted with a circle. Joint angles are limited to the following bounds: shoulder flexion-extension $\theta_1 \in [-1.0, 2.9]$, abduction-adduction $\theta_2 \in [-1.5, 1.5]$, internal-external rotation $\theta_3 \in [-1.0, 1.0]$, elbow flexion-extension $\theta_4 \in [0.0, 2.8]$, pronation-supination $\theta_5 \in [-0.7, 1.0]$, wrist abduction-adduction $\theta_6 \in [-1.5, 1.5]$, flexion-extension $\theta_7 \in [-0.5, 0.7]$. With $\theta_i = 0$ for all joints the arm is stretched and pointing to the right.

8.3 Simulation and Tasks

As a proof of concept, the complete framework is tested with a minimalistic robot arm simulation. This allows to compare the three considered FA algorithms before advanced issues such as noise, time delays, or dynamics come into play. Since brain functionality shall be modeled, a 7 DoF anthropomorphic arm is simulated with dimensions roughly comparable to a human arm. The simulated arm is composed of three limbs as illustrated in Figure 8.3. For each joint the maximum rotation velocity is restricted to 0.1 radians per simulation step. Simple reaching (pointing) tasks are considered and therefore the three-dimensional Cartesian space of hand locations defines the task space.

8.3.1 Target Generation

Training or testing targets $\xi^* \in \Xi$ cannot be generated completely random because a) the task may provoke a deadlock (cf. Section 7.4) or b) a random task location can be near the current location, which makes the task too easy. Both facts falsify performance measurements. Instead, tasks should be generated such that all of them can be accomplished with a simple directional control scheme (Section 7.4) but require at least a few simulation steps.

Ideally, the tasks are distributed uniformly random in the control space to ensure that the full forward kinematics can be learned during training and that the full mapping is also tested when performance is measured. An exception are the joint limits, which should be avoided because the Jacobian cannot represent the limits. Contrary to a uniform sampling, animals and humans train sub-spaces on demand, but do not train every posture equally well [37]. However, this would require a *weighted* performance assessment depending on the configuration, which seems overly complicated. Thus, a uniform sampling of the inner region of the control space seems favorable.

The following method for task generation satisfies the above requirements. The control space under consideration for training targets is reduced by a certain *border fraction* (typically 10% on each side). Suppose the current configuration is $\boldsymbol{\theta}_0$ with a corresponding task space location $\boldsymbol{\xi}_0$. The indices used here are not to be confused with individual joint angles or task space dimensions. A random control space *direction* $\Delta\boldsymbol{\theta}$ is generated and its length scaled down to a sufficiently small value, e.g. one percent of the smallest joint range. The current configuration $\boldsymbol{\theta}$ is incremented iteratively in the given direction

$$\boldsymbol{\theta}_k = \boldsymbol{\theta}_{k-1} + \Delta\boldsymbol{\theta} \tag{8.4}$$

until the task space difference $\Delta\boldsymbol{\xi} = \boldsymbol{\xi}_k - \boldsymbol{\xi}_0$ is larger than some threshold $\Delta\boldsymbol{\xi}_{\min}$. The resulting robot task is then comprised of a starting control space configuration $\boldsymbol{\theta}_0$ and a task space target $\boldsymbol{\xi}_k$, while intermediate steps $1, \ldots, k-1$ are ignored. The process is repeated with the generated direction until a joint limit is hit. Then, a new control space direction is generated. Figure 8.4 shows a five-step trajectory for a simple, planar 2 DoF arm. The resulting four tasks are given as (start-posture, target-location) pairs: $(\boldsymbol{\theta}_1, \boldsymbol{\xi}_2)$, $(\boldsymbol{\theta}_2, \boldsymbol{\xi}_3)$, $(\boldsymbol{\theta}_3, \boldsymbol{\xi}_4)$, and $(\boldsymbol{\theta}_4, \boldsymbol{\xi}_5)$.

This method provides tasks with a comparable task space difference clearly defined by the user, avoids deadlocks by separation into small movements, and samples the control space thoroughly while the joint limits are avoided. This does not mean that the outer region is never sampled during *control*. If the trained model tells the controller to go towards the joint limit, samples generated there are trained as well. By putting initial and final postures apart from the boundaries simply helps to avoid deadlocks and gives more meaning to the performance measurements, because those neither respect joint limits.

Finally, tasks for training and test phases are generated independently with a fixed set of 100 tasks for testing, while as many training tasks are generated on demand as required for training during an experiment.

8.4 Evaluating Model Performance

It is not immediately clear how performance of a robot controller should be evaluated: The control space is large and one can hardly evaluate every posture

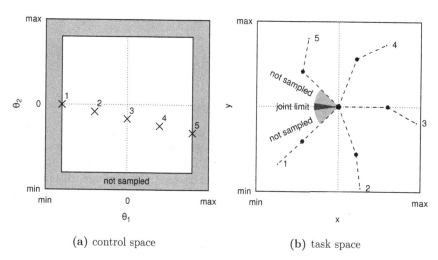

(a) control space (b) task space

Figure 8.4: Tasks for a two-joint planar robot arm are generated along a trajectory through the control space until a joint limit is reached. (a) The control space is reduced by 10% on each side, such that the outer region is not sampled. Five configurations, labeled 1 to 5, are sampled along a fixed direction until a boundary is hit. (b) The corresponding task space locations 1 to 5 are sufficiently far apart from each other to make the task difficult, but the distance is small enough to not provoke a deadlock. The first joint ranges from $-172°$ to $+172°$, which leaves a $16°$ blocked area. The sampled joint range is further reduced to $[-137.6°, +137.6°]$.

and each movement direction. For a single task, from initial posture to target, a varying number of steps is required and multiple objectives may be desired: fast convergence, high precision, and low energy consumption, to name a few. Additionally, precision and speed are somewhat conflicting. Therefore multiple measures are computed to get a complete picture of the model quality.

Goal Reached? A target is considered as *successfully reached*, when the distance between hand and goal is less than 5% of the arm's length. After a sufficiently large number of simulation steps (roughly 100 times the steps required for a typical reaching movement) a failure is recorded and the next task is evaluated.

Number of Steps required for one task. This basic measure is related to the convergence speed, which depends on the correctness of the model but also on the actual task space distance to be traveled.

Path Efficiency is a measurement of *straightness* of the task space path from start to end. If the goal is not reached the measure evaluates to zero.

Otherwise, for a movement consisting of k simulation steps, a value in $(0, 1]$ is computed as

$$\frac{\|\boldsymbol{\xi}_k - \boldsymbol{\xi}_1\|}{\sum_{i=2}^{k} \|\boldsymbol{\xi}_i - \boldsymbol{\xi}_{i-1}\|} \tag{8.5}$$

which evaluates to one, if the task space path is perfectly straight.

Angular Error is measured as the angular deviation of the *actual* direction $\boldsymbol{\xi}_i - \boldsymbol{\xi}_{i-1}$ from the *optimal* direction $\boldsymbol{\xi}^* - \boldsymbol{\xi}_{i-1}$ towards the target. In contrast to the above measures, this one is evaluated for every simulation step i individually, independent of the number of steps. Furthermore, this measure makes sense even if the goal is not reached. For a movement consisting of k steps, an average angular error is computed as

$$\frac{1}{k} \sum_{i=1}^{k} \arccos \left(\frac{(\boldsymbol{\xi}^* - \boldsymbol{\xi}_{i-1})^T (\boldsymbol{\xi}_i - \boldsymbol{\xi}_{i-1})}{\|\boldsymbol{\xi}^* - \boldsymbol{\xi}_{i-1}\| \|\boldsymbol{\xi}_i - \boldsymbol{\xi}_{i-1}\|} \right), \tag{8.6}$$

where the geometric meaning $\boldsymbol{a}^T \boldsymbol{b} = \cos(\gamma) \|\boldsymbol{a}\| \|\boldsymbol{b}\|$ of the dot product is exploited. A division by zero must be avoided, when the robot device is not moving, or once the target is reached exactly. In that case, the angular error is not evaluated.

8.5 Experiments

Previous work [14, 19, 84, 83] has proven that the outlined framework works with the XCSF algorithm and the forward kinematics of a 7 DoF anthropomorphic arm can be successfully learned for accurate, inverse control. The LWPR algorithm was used to learn forward kinematics of anthropomorphic arms as well [74].

While XCSF and LWPR are compared in [78, 25, 24], a detailed performance evaluation is missing. Furthermore, the number of RFs was different for each algorithm and that allowed LWPR (theoretically) to reach a better precision. Notably, [24] shows for the first time that XCSF and LWPR are able to learn a kinematic model of a *real* robot from *scratch*. Unfortunately, a valid performance measurement is almost impossible on a real robot, because tasks are not reproducible.

The aim of this thesis is a thorough and fair comparison in a tractable scenario. Therefore XCSF, LWPR, and additionally RBFN are compared directly on the same robotic device, the same task space targets, and with the same controller. Those algorithms are difficult to compare directly, as they inherently share a population size versus accuracy trade-off: Decreasing the number of RFs, results in a reduced accuracy as well.

Here the aim is to give roughly the same amount of RFs to each algorithm, although that cannot be ensured exactly: LWPR produces RFs on demand and one can only tune parameters indirectly to influence the population size. XCSF on the other hand has a fixed maximum population size. Due to the evolutionary overhead, however, this value is less meaningful and the *final population size after condensation* should be emphasized instead. Finally, RBFN have a fixed population size, but have no means to distribute the RFs throughout the input space. Thus, the population size depends on the initialization, which could be a uniform grid covering the full input space.

To prove that a non-linear model is required at all, the first experiment uses as single linear model to learn the full forward kinematics of the 7 DoF anthropomorphic arm. In turn, the three algorithms RBFN, LWPR, and XCSF are applied as well.

8.5.1 Linear Regression for Control

On the one hand, it is known that the forward kinematics of rotational joints are represented in a sinusoidal wave function. On the other hand, that function *could* be comparably flat and a linear model *could* achieve an acceptable accuracy for kinematic control. Obviously this is not the case, but evaluating a linear model allows for some important insights.

Therefore a single linear RLS model is trained to approximated the 7 DoF velocity kinematics as detailed in Section 8.1. The input space is the seven-dimensional joint velocity space $\dot{\Theta}$, while the predicted output is the resulting task space velocity $\dot{\xi}$, that is, the 3D velocity of the end effector. First, the simulation is run for 50000 (denoted as 50 k) steps, while the simple controller from Section 8.2 tries to reach the generated task space targets. When the distance between end effector and target is less than 5% of the kinematic chain's length, the next target is activated. In each iteration, the linear model is updated with the velocity data from the previous iteration. At the same time the model is used to provide the Jacobian matrix for inverse control (Section 7.4).

Initially this is a dilemma: who came first, the model or the movement? Without a movement, the model cannot learn, and without a model, the controller cannot generate a movement command. However, model initialization with external data (often termed "motor babbling") is *not* required, but instead a small amount of motor noise, gravity, or any other external force is sufficient to initialize the model *somehow*. The model is probably wrong for the first steps, but it is then able to produce a non-zero Jacobian, which in turn results in *some* movement and the whole learning process gets started. Trying to reach targets right from the start, without a motor babbling phase, is termed "goal babbling" [70].

As mentioned earlier, it is important to distinguish *training* and *offline* performance: When the model is trained during control (also termed online learning), the model is updated with samples from the current trajectory such that the previous movement is the learning input for the current iteration. The subsequent Jacobian is always similar to the current one, and even if the model is one step behind, its performance is probably sufficient for control. When learning is disabled during a certain test phase, the model knowledge about the forward kinematics is tested.

During the 50 k training iterations, the number of successfully reached goals is recorded amongst the other performance statistics. In turn, the learning is disabled, that is, Jacobians are requested for control but the local models do not learn from the resulting movements. 100 more reaching tasks are evaluated to measure the true knowledge of the model. As always, 20 independent runs are made to compute minimum, maximum, and median performance along with the 25% and 75% quartiles.

Figure 8.5 shows intriguing results: During training, more than 80% of the goals are reached with the single linear model. While this could be explained by the linear model following the trajectory, even during the offline test phase around 70% goals are successfully hit. However, taking a closer look at the other performance measures, the actual reaching trajectories are rather bad. On average, the angular error is about 50° during training and 55° during testing.

Analyzing individual tasks and respective trajectories shows that the linear model does well at the center of the task space, as the Jacobian is comparably accurate due to averaging. At the outer region, however, the controller starts spiraling in a elliptic fashion around the target, because the model provides wrong directions.

The experiment has shown that a linear model is not sufficient to accurately represent the forward kinematics. More importantly however, we now have a feeling for the performance measures of a comparably bad controller. First, reaching about 70% of the targets is not difficult and it takes less than 25 steps on average. However, an angular error of about 50° is not acceptable and the corresponding path efficiency is about 60%, that is, the actual trajectory is roughly 1, 66 times longer than a straight path.

Another insight gained from the last experiment is the difference between online learning and offline testing. The RLS model is able to slightly adapt to the current trajectory resulting in an overall improved online performance. However, since the model aims to include all data points seen so far (no forgetting), the model stabilizes mostly. If the forget rate is set to a low value, the RLS model

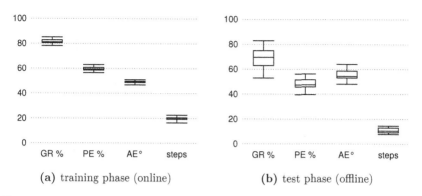

(a) training phase (online) (b) test phase (offline)

Figure 8.5: Performance of a single linear RLS model controlling the 7 DoF anthropomorphic arm. Box and whisker plots show the minimum and maximum performance as whiskers and the inner 50% of the measurements lies inside the box. The median is drawn as a line in the box. (a) Performance measures during online training from left to right: percentage of goals reached (GR), average path efficiency (PE) as a percentage of straightness, average angular error (AE), and average number of steps to the target. (b) Final offline performance, where learning is deactivated.

aggressively follows the current trajectory and quickly loses track of the samples further apart.

Figure 8.6a shows the online performance for a RLS model with a forget rate $\lambda = 0.6$. At least 99.9% goals are reached with about 96% path efficiency during *training* in all runs and overall the performance is stable – the box and whiskers become a single line in the plot. The angular error is reduced to about 15° and less than 5 steps suffice to reach a target on average during training.

Obviously, when learning is disabled the model remains at the Jacobian which was trained last and, thus, offline performance is more or less random (Figure 8.6b). Eventually training ends at a centered configuration, where performance is comparable to a static linear model but often a configuration further towards a joint limit is stored, where performance quickly degrades during offline testing.

The experiment confirms that a single linear model allows for accurate control when the forget rate is adequate and the control frequency is large enough[2]. Despite the good performance, such a rapidly adapting model can not properly explain brain function as long term learning effects (e.g. athletes that train

---

[2]The higher the frequency of model updates, the closer is the current configuration to the previous, learned one.

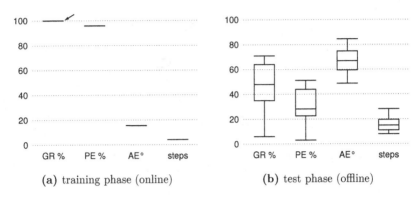

(a) training phase (online)   (b) test phase (offline)

Figure 8.6: A linear RLS model with forgetting is able to consistently adapt to the current trajectory, when learning is activated. (a) Almost all goals are reached (99.9% at least) and the box and whiskers collapse into a line, which is highlighted with an arrow. (b) On the contrary, the offline performance is poor.

a specific movement over and over) are not possible. Therefore the non-linear function approximators are now evaluated.

8.5.2 RBFN

The considered kernel regression methods RBFN, XCSF, and LWPR all share a trade-off between accuracy and population size. The more RFs are available to one such algorithm, the higher precision can be obtained. One way to get a fair comparison is a exhaustive search along the Pareto front for those two objectives as done in [87]. However, the computational time required to get a sufficiently fine grained Pareto front becomes unacceptably large in the present context: Each algorithm must run multiple times for a single data point to average the noisy results. It is not clear what the optimal settings are for any given algorithm, thus, many tangential parameters have to be optimized – resulting in even more evaluations. Here the full kinematic simulation has to be carried along additionally. Finally, it may result in one algorithm being better in one part of the Pareto front and the other one being superior elsewhere, which ultimately does not provide a clear ranking [87]. Thus, a simpler route is taken here: The algorithms are set up such that the number of RFs is roughly the same, then other parameters are tuned to the given problem, and finally the resulting control performance can be compared on a fair level.

For a RBFN, the number of RFs depends on user preference. Either, RFs are placed onto random data points, or distributed manually throughout the input space. The latter choice seems more appropriate to the task of learning forward kinematics, where a uniform distribution is a best guess to start with. All RFs are circular and share the same radius. Thus, it presents a good distribution as the overlap is minimized as opposed to XCSF, where evolution produces a highly overlapping population of RFs. On the other hand, ellipsoidal shapes with individually sized RFs may be beneficial for the problem hat hand – this question is answered in the subsequent section.

With a seven dimensional control space for the 7 DoF anthropomorphic arm, a uniform distribution of the neuron population requires k^7 RFs. With $k = 1$ there is only one neuron which resembles the case of a single linear model. Having $k = 2$ RFs per dimension still seems very sparse. On the other hand, $4^7 = 16384$ is far more than previous studies required for successful control [14, 19, 84, 83, 78, 25, 24], where the number of RFs ranged from 500 to 4000.

Therefore $3^7 = 2187$ neurons with radius $\sigma = 0.25$ are distributed uniformly over the control space in the following experiment. Each neuron of the RBFN is equipped with a linear RLS model. Online training and offline testing phases alternate, where each training phase consists of 1000 simulation steps and each testing phase evaluates 100 task space targets. The network is trained for a total of 200 k iterations. However, now the focus lies on the offline tests that measure the true knowledge as opposed to the dynamic trajectory tracking during online training. Twenty independent runs are made, although the RBFN approach shows minimal variance.

The control performance shown in Figure 8.7 reveals that the RBFN rapidly learns an accurate mapping in less than 20 k iterations. After about 8 k iterations all goals are reached reliably. The angular error stays at about 20.5° with a sufficient path efficiency of about 98.2%.

To summarize briefly, the comparably simple RBFN approach is able to quickly learn an accurate representation of the forward kinematics of an anthropomorphic 7 DoF arm in a 3 D task space. Performance is stable as location, shape, and size of the RFs are fixed. Thus, each RLS model has a fixed region to approximate and converges quickly.

8.5.3 XCSF

Compared to an RBFN, the XCSF algorithm has more flexibility in its solution expressiveness, but also requires more parameters to be tuned. Instead of a spherical RF structure, an ellipsoidal one is used now. An RF is still initialized as a sphere, but the GA in XCSF may then optimize the shape further. The initial radius of RFs was set to $\sigma = 0.3$ which is slightly larger than for the RBFN, because XCSF a) inherently produces an overlapping population, b) does not use the Gaussian activation function and thus its RFs have a much smaller

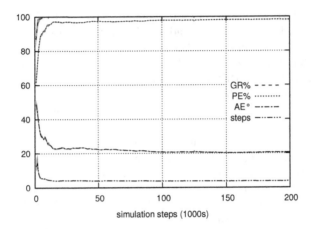

simulation steps (1000s)

Figure 8.7: Control performance during offline tests of a RBFN with 2187 neurons distributed uniformly over the control space.

area of influence, and c) must still be able to cover the whole input space to avoid a covering-deletion cycle.

The frequency of the GA is controlled by the θ_{GA} parameter, which, by default, would be set to 50. However, as the robot arm moves along a trajectory, it takes longer to properly estimate the true fitness of an RF. Thus, θ_{GA} is set to 200 to provide a more reliable fitness to the GA as done in other related works [85, 83]. For the same reason, the RLS model of a newly created RF requires some iterations to adapt to the new region. Thus, RFs that have seen less than $2n = 14$ samples are excluded from control [83].

Regarding the population size, some manual experimentation has shown that a maximum population size of 5500 can be reduced by condensation to less than 2064 RFs[3]. Although this is considerably less than RBFN's number of neurons, the control performance is comparable. After 20 independent runs, the median population size was 1856.5 RFs after condensation, the minimum was at 1727, and the maximum was 2064. The median is approximately 15% below the population size of the RBFN.

The kinematic control performance of XCSF is shown in Figure 8.8. The first observation is that XCSF's control is not as stable as a RBFN. Since evolution may produce some improper RFs from time to time, and the respective RLS models have to adapt to the new region, control is slightly noisy. However, once

---

[3]Other parameters were set to defaults: $\beta = 0.1$, $\nu = 5$, $\xi = 1$, $\mu = 1/35$, $\tau = 0.4$, $\theta_{del} = 20$, $\theta_{sub} = 20$. Condensation starts at 95% of the learning time.

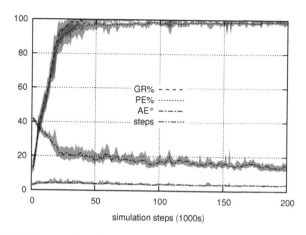

simulation steps (1000s)

Figure 8.8: XCSF's control performance during offline tests for the 7 DoF anthropomorphic arm. After about 40 k iterations, a suitable model of the forward kinematics is built.

evolution has found a suitable distribution of RFs at 50 k iterations, all goals are reached in 75% (15 out of 20) of the runs.

More importantly however, the angular error is lower than for a RBFN despite the lower population size. With XCSF the angular error converges to about 14.7° (best case 12.4°, at worst 16.5°) with a path efficiency of about 98.5%, whereas RBFN achieved an angular error of about 20.5° (best case 20.4°, worst 20.7°). Thus, XCSF was able to find a slightly better clustering of the control space compared to the uniformly distributed spheres in the RBFN scenario.

8.5.4 LWPR

Finally, LWPR is tested as a learning algorithm in the kinematic control framework. Again, there is no direct way to specify the final population size. Instead, the initial size of RFs is the main parameter to influence the final population size, as new RFs are only created when no other RF has a sufficient activation for a given sample. Setting $\texttt{init\_D} = 27$, which is the squared inverse radius of the Gaussian kernel ($\sigma = 1/\sqrt{27} \approx 0.2$), results in a fitting population size. In 20 independent runs, the median number of RFs was 2159 (minimum 2109, maximum 2220), which closely resembles the population size of the RBFN. Manual experimentation with other parameters has shown that the default values yield the best results in terms of kinematic control performance[4].

---

[4]Other evaluated parameters were left at their respective defaults: $\texttt{w\_gen} = 0.1$, $\texttt{w\_prune} = 1.0$, $\texttt{penalty} = 10^{-6}$, $\texttt{update\_D} = 1$, $\texttt{meta} = 0$.

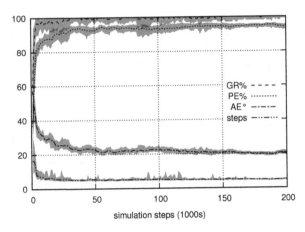

Figure 8.9: LWPR's control performance during offline tests.

Figure 8.9 confirms that LWPR is as well able to learn an accurate representation of the seven dimensional anthropomorphic forward kinematics. More than 99% of the goals are reached reliably with a path efficiency of about 94.5% (worst case 92.5%, best case 95.6%) and an angular error of 20.2° (minimum 19.2°, maximum 21.7°).

8.5.5 Exploiting Redundancy: The Effect of Secondary Constraints

Up to now, it was only tested *if* goals can be reached and how *accurate* that process is. However, the task space is only three dimensional while the control space is highly redundant with seven DoF. Thus, it would be sufficient to learn the most relevant joints (e.g. shoulder and elbow only) for all configurations instead of the full kinematic forward mapping. More precisely, up to now it would be fine to ignore the nullspace. Instead it was shown that the learned model still represents redundant alternatives sufficiently [83]. The experiment is briefly replicated within the present framework.

After training for 30 k simulation steps, learning is disabled for a redundancy test. All joint angles are set to zero which leaves the arm in a fully stretched posture pointing straight away from the body. The corresponding task space location is $\boldsymbol{\xi} = (100\,\mathrm{cm}, 0\,\mathrm{cm}, 0\,\mathrm{cm})^T$. The primary reaching target is $\boldsymbol{\xi}^* = (25\,\mathrm{cm}, 0\,\mathrm{cm}, -70\,\mathrm{cm})^T$, which is a location centered in front of the (imaginary) body and thus leaves enough freedom in control space for redundant alternatives.

During the first 100 simulation steps of the redundancy test the default constraint is used: A centered configuration $\boldsymbol{\theta}_{\mathrm{center}} = (0.95, 0, 0, 1.4, 0.15, 0, 0.1)^T$

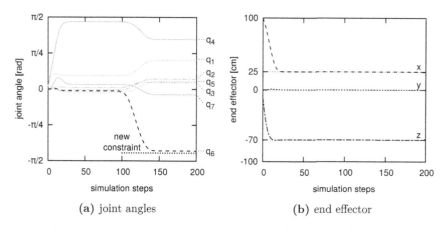

(a) joint angles   (b) end effector

Figure 8.10: Changing the secondary constraint confirms that redundancy is represented in the learned model. (a) For the first 100 simulation steps a centered configuration is preferred. Thereafter, the constraint is changed for joint θ_6 whose preferred angle is now -1.4 radians. The controller is able to almost fully realize the new constraint at about simulation step 140. (b) For the full duration, the primary target is $(25\,\mathrm{cm}, 0\,\mathrm{cm}, -70\,\mathrm{cm})^T$, which is reached after less than 20 steps. Changing the secondary constraint does not affect the task space location.

is preferred and joint limits are avoided. For the current configuration θ this is realized by setting

$$\Delta q_0 = \theta_{\mathrm{center}} - \theta, \qquad (8.7)$$

which results in a joint angle *displacement* towards a centered configuration. The vector Δq_0 is then scaled to a reasonable *velocity* $\dot{q}_0$ to be put into Equation (7.17).

After 100 steps, the constraint is changed for the wrist joint θ_6, which shall now aim for an angle of $\theta_6 = -1.4$ (the limit is -1.5 radians). With the previous "centering" constraint, the preferred angle was $\theta_6 = 0$. The resulting trajectory is shown in Figure 8.10 for the XCSF learner, confirming that redundancy is well represented in the learned model. Results for the other learners are almost identical and are not reported.

8.5.6 Representational Independence

The kinematic mapping was characterized by joint angles in the control space that were mapped to the *Cartesian* location $\xi = (x, y, z)^T$ of the end effector. However, the framework does not rely on this particular representation. Instead, any other relation could be learned, as long as the control space configuration uniquely defines the task space location.

However, a different mapping will generally yield a different performance as the model representation (kernel shape, type of local model, weighting strategy) interacts with the structure of the mapping to be learned (cf.Section 3.4). When locally linear models are employed to approximate subspaces, the overall linearity of the kinematic mapping has a strong impact.

To evaluate the effect of a different representation, the former Cartesian encoding of the task space location is replaced by an *egocentric* encoding consisting of the horizontal angle, the vertical angle, and a distance measure [83]. The viewpoint of the imaginary vision system is placed $15\,\mathrm{cm}$ to the left of and $10\,\mathrm{cm}$ above the shoulder. Given the Cartesian task space coordinates $(x, y, z)^T$ with respect to the egocentric viewpoint, the new task space coordinates are computed as[5]

$$\varrho = \mathrm{atan2}(x, -z)\,,$$

$$\psi = \mathrm{sign}(y)\arccos\left(\frac{x^2 + z^2}{d\sqrt{x^2 + z^2}}\right)\,,$$

$$d = \sqrt{x^2 + y^2 + z^2}\,, \tag{8.8}$$

where the horizontal angle $\varrho \in [-\pi, \pi]$ is defined in the x-z-plane and a value of zero indicates a view to the front, while the vertical angle $\psi \in [-\pi/2, \pi/2]$ is measured between $(x, y, z)^T$ and $(x, 0, z)^T$. When the task space location coincides with the center of view, that is, $x = y = z = 0$, both angles are undefined and set to zero. For $x = z = 0$ and $y \neq 0$, that is, the end effector is located exactly above or below the center of view, the horizontal angle is zero while ψ is set to $\mathrm{sign}(y)\pi/2$.

This representation contains a point gap for both angles at the origin and another gap for the horizontal angle, when the end effector passes through the vertical line at $x = z = 0$. On the other hand, the relation from joint angles to the new task space representation is *more* linear. For example, two shoulder joints are almost linearly correlated to the vertical and horizontal angle, while distance is not affected by those joints. This allows for strong generalization, that is, RFs can extend over the shoulder angle dimensions. This allows to use more RF resources for the remaining joints. Altogether, an improvement in control performance can be expected, while control may fail for some tasks close to the representational gaps.

---

[5]$\mathrm{atan2}(x, -z)$ evaluates to $\arctan(-x/z)$ for $z < 0$, $\arctan(-x/z) + \pi$ for $z > 0 \wedge x \geq 0$, $\arctan(-x/z) - \pi$ for $z > 0 \wedge x < 0$, $\pi/2$ for $z = 0 \wedge x > 0$, $-\pi/2$ for $z = 0 \wedge x < 0$, and 0 for $z = x = 0$.

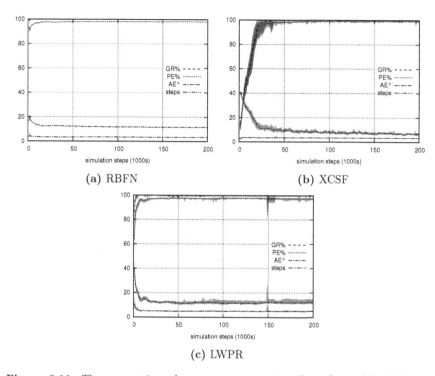

(a) RBFN   (b) XCSF

(c) LWPR

Figure 8.11: The egocentric task space representation allows for a slightly better control performance compared to the Cartesian representation.

The actual performance for RBFN, XCSF, and LWPR in Figure 8.11 indeed reveals that the overall angular error is lower for all algorithms. The parameter settings from the previous experiments were used. The RBFN with 2187 RFs achieves a median angular error of 11.8° and a 98.1% path efficiency. XCSF evolved about 1790 RFs (minimum 1709, maximum 1976), but also achieved a very low angular error of 6.8° (minimum 6.1°, maximum 7.5°) and a 99% path efficiency. Thus, XCSF outperforms the RBFN in both population size and accuracy. With the new representation, the variability in LWPR's population size is comparably large: the minimum number is 1784, median 1984, and maximum at 2123. A 97% median path efficiency is achieved with an angular error of 12.4° (minimum 10.8°, maximum 14.9°). Again, XCSF achieved a better control performance with fewer RFs.

The experiment has illustrated that the learning and control framework is flexible in terms of the actual representation, but also revealed that the actual performance may change considerably for a different kinematic mapping.

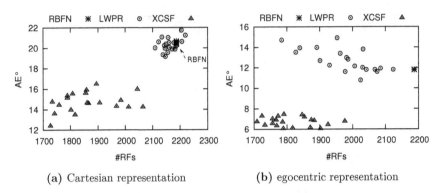

(a) Cartesian representation   (b) egocentric representation

Figure 8.12: A multi-objective view on the kinematic control performance of RBFN, LWPR, and XCSF. The two objectives (angular error and population size) are conflicting and both to be minimized. XCSF outperforms the other algorithms in both objectives, but only a small niche of the parameter space was tested for all algorithms.

8.6 Summary and Conclusion

In this chapter, all pieces were put together for a successful kinematic control framework for an anthropomorphic 7 DoF arm in a 3 D task space. The first experiment has shown that a single linear model is not sufficient to accurately represent the forward kinematics. However, three neuron-based approaches with multiple locally linear models (RBFN, XCSF, and LWPR) were able to learn an accurate mapping of the kinematics for reliable control.

As a final comparison of the three considered learning algorithms, the results in the two major objectives – angular error and population size – are plotted against each other in Figure 8.12 for both evaluated task space representations. On the one hand, the graphs show that XCSF outperforms the simple RBFN approach and LWPR in both objectives. On the other hand, only a very small part of the parameter space was evaluated for all algorithms and, thus, conclusions drawn from this graph should be taken with a pinch of salt. Instead of comparing the algorithms in depth, the conclusion is that all algorithms were able to learn a sufficient forward kinematics model for accurate control, but probably none of them *exactly* resembles brain functionality.

However, the framework fulfills all requirements stated in the early Section 1.4. A model is learned *online*, that is, training and execution take place simultaneously. Furthermore, the model is not initialized with some external knowledge but trained from scratch. No restrictions with respect to control space or task space representation is made: Humans may not necessarily have a Cartesian coordinate system and the outlined framework does not depend on it either, which was also confirmed experimentally. Moreover, redundancy is not resolved but learned and it was shown that redundant movement alternatives can be exploited flexibly during control.

Finally, the three algorithms RBFN, LWPR, and XCSF can be seen as neural networks that learn a directional encoding[6]. Although there is no proof that the brain is structured exactly this way, the overall framework may be seen as a semi-plausible implementation or explanation of *brain functionality* with respect to learning body control.

---

[6]Implementation-wise LWPR does not learn a directional, that is, first order velocity mapping but the zeroth order mapping. Probably this algorithm could be modified accordingly, but this is beyond the scope of this thesis.

9 Visual Servoing for the iCub

The learning and control framework was applied to a purely kinematic simulation where neither external forces, inertia, nor noise disturb the learning process. The present chapter introduces a more realistic scenario, where a physics engine complements the simulation and the end effector location is not simply *available*, but is *sensed* by means of stereo cameras. This is also called *visual servoing* [21], where vision is used for closed loop control.

The iCub [69] simulation is used for the next experiments, where only XCSF – which came up with the most accurate model in previous experiments – is applied as a learner. The iCub is an anthropomorphic robot built as a 3.5 year old child. It features 53 servo motors for head, arms, hands, torso, and legs as well as two cameras installed in the head. Every joint has proprioceptive sensors that measure joint angle and velocity. Many other features such as a sense of touch are available but not relevant for this work. Figure 9.1 shows real and simulated iCub.

When the *real* vision through stereo cameras is used to sense the end effector location, the learning and control framework must be extended. First, the end effector must be visible on the camera images and consequently the head must move such that target and end effector are visible. Therefore the forward kinematics of the head have to be learned as well. Second, the 3 D end effector location must be computed from stereo 2 D pixel images. Third, the movement of arm and head both influence the location of the end effector on the camera images and this interdependence must be disentangled. This chapter replicates experiments done elsewhere [78, 24] and adds the performance measures used before in Chapter 8.

Section 9.1 explains how the cameras are used to sense the task space location and discusses related coordinate systems. Section 9.2 in turn gives details about the kinematic mapping to be learned and briefly explains necessary modifications of the control framework. Finally, experimental results are presented in Section 9.3, where the iCub draws an asterisk shape after sufficient training.

9.1 Vision Defines the Task Space

As before the tasks are simple reaching tasks, where iCub's hand shall move to a given target. However, now the stereo cameras are used to sense the task

(a) real robot   (b) simulation

Figure 9.1: Images of the iCub robot: (a) shows the real robot during a grasping demonstration. (Permission to reproduce the image was granted by the Robotcub Consortium, www.robotcub.org.) (b) The simulation replicates the real robot and its physical properties.

space. Both cameras produce a steam of 320×240 pixel images[1]. Extracting the location of the end effector is a complex problem beyond the scope of this thesis. Therefore a clearly visible green ball is attached to the hand, which is distinct from everything else in the scene. Computing the location of the ball in a 2 D image is comparably simple, especially in a noise free simulation, where simple thresholding is sufficient to extract relevant pixels. But what is *location* here actually?

A single camera image is a two-dimensional pixel based coordinate system without depth. Combining left and right camera images allows to compute the depth of an object also known as *reprojection* [79, section 22.1.2], which will be detailed later. This step converts the 2 D pixel location into the 3 D "world", where the unit becomes meters. However, the frame of reference is the head, not the torso or some fixed point in the world itself. Thus, *location* depends on the frame of reference. Although the learning framework does not require knowledge about any of those coordinate systems, it helps to better understand the interplay between arm movements, head movements, and the location of the end effector (Figure 9.2).

For reaching tasks as considered here it is convenient to use the *head* reference frame as the task space for several reasons. First, the unit is meters and thus precision is not defined in pixels. Second, the head frame is centered between the two cameras which simplifies the tracking of the ball by maintaining a centered ball location in that frame of reference. Most importantly, however, it is

---

[1]Other versions of the iCub may have a different resolution.

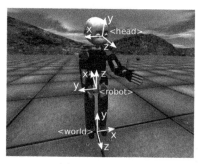
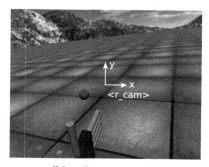

(a) external view   (b) right camera view

Figure 9.2: Detecting the end effector location by means of vision is a complex problem, which is here simplified by attaching a green ball to the left hand. The connection between hand and ball is not visible, but simulated. (a) Three relevant coordinate systems are visualized: The *world* coordinate system has its origin on the floor. The *robot* coordinate system is centered on the torso pitch rotation axis. Since legs and torso are deactivated in all experiments, the transformation from world to robot is a fixed roto-translation. The *head* coordinate system, however, depends on the joint angles of the neck. (b) An image obtained from the right camera. The corresponding coordinate system is related to the *head* frame of reference, but units are pixels and there is no depth information. Since the other camera is mounted a few centimeters to the left, the images are different.

sufficient to learn a *single* kinematic model by considering the kinematic chain from end effector to head including the joints of the arm and the neck. Before the learning framework is detailed, a brief introduction to the reconstruction of 3 D location from stereo cameras is given.

9.1.1 Reprojection with Stereo Cameras

Given two camera images that both contain a certain object, one can reproduce its three-dimensional location in space, when the *extrinsic* and *intrinsic* camera parameters are known. Extrinsic camera parameters are the location and angle in the world, while intrinsic parameters refer to the conversion into pixels: focal length and size of the resulting image. When a camera takes a picture of an object, the object is *projected* onto the image. Here, the reverse is needed – reconstruction of object location from an image. Since depth information is lost during the projection (assume that object size is unknown), a second camera image from a different location is required to *triangulate* the original object location, which is essentially a geometric problem. This process is also called reprojection.

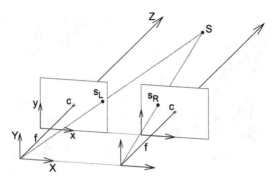

Figure 9.3: Geometry of stereo cameras. Camera intrinsics are labeled with lowercase letters, extrinsics with uppercase letters. Each camera has a global 3 D coordinate system (denoted with X, Y, Z for the left camera) with its Z axis going through the center c of the image (vergence is not used). The distance from focal point (origin) to image plane is termed focal length f. The X-axes are aligned. The 2 D image plane is the pixel based internal coordinate system. Again, x-axes are aligned to produce rectified images. A point S in the world is projected onto the image planes resulting in pixel locations s_L and s_R for left and right image respectively. The dashed projection lines from point S to the camera focal points are called epipolar lines.

While it is possible to do this with arbitrary cameras at different angles, with different resolutions, the whole process becomes particularly simple with identical stereo cameras that both face the same direction as built in the iCub. Images obtained from such cameras are said to be *rectified*, which is illustrated in Figure 9.3 with *pinhole* camera models.

The projection of a point $S = (S_X, S_Y, S_Z)^T$ onto the image plane of a camera can be computed by a matrix multiplication in homogeneous coordinates

$$\begin{pmatrix} s_x \\ s_y \\ 1 \end{pmatrix} \sim \underbrace{\begin{pmatrix} f_x & 0 & c_x & -f_x T_X \\ 0 & f_y & c_y & 0 \\ 0 & 0 & 1 & 0 \end{pmatrix}}_{\text{camera matrix}} \begin{pmatrix} S_X \\ S_Y \\ S_Z \\ 1 \end{pmatrix}, \tag{9.1}$$

where the $\sim$ sign denotes equality up to a scale factor, f_x and f_y are the focal length parameters of the camera, $(c_x, c_y)^T = c$ is the principal point (often the center of the image), and T_X is an external translation parameter used to align left and right camera coordinate systems. The value can be chosen such that the right camera coordinate system is translated to the origin of the left camera system by using the physical distance between the cameras. Correspondingly, T_X is set to zero for the left camera[2].

The only difference between iCub's left and right camera matrices is the external translation parameter T_X. Thus, the resulting projections only differ in the x-coordinate. This particular difference is the disparity d used to reconstruct the depth information from two images. Again, this can be formulated by a homogeneous matrix multiplication

$$\begin{pmatrix} P_X \\ P_Y \\ P_Z \\ W \end{pmatrix} = \underbrace{\begin{pmatrix} 1 & 0 & 0 & -c_x \\ 0 & 1 & 0 & -c_y \\ 0 & 0 & 0 & f_x \\ 0 & 0 & -1/T_X & 0 \end{pmatrix}}_{\text{reprojection matrix}} \begin{pmatrix} s_x \\ s_y \\ d \\ 1 \end{pmatrix} , \qquad (9.2)$$

where $S' = (P_X/W, P_Y/W, P_Z/W)^T$ are the reconstructed scene coordinates. Here the point $(c_x, c_y)^T$ from the left camera is reprojected, while the disparity d between left and right image point is fed into the third column. The physical distance T_X between the cameras completes the triangulation problem. The point S' is represented in the left camera's coordinate system, but a translation by $T_X/2$ in X direction yields the *head* reference frame where the origin is centered between the cameras. Alternatively, that translation can be directly put into the above equations, however, at a loss of readability.

The object to be projected (and reprojected) is the green sphere attached to the hand as illustrated in Figure 9.2. Given the two camera images, a simple green threshold yields the location in pixel coordinates. No noise is present in simulation and advanced processing is not required for this step. Next, the reprojection from Equation (9.2) yields the location of the sphere in the *head* coordinate system, which is the task space for all iCub experiments. Although the green sphere does not exactly represent the location of the hand, it is sufficient for the purpose of learning a kinematic forward model of arm and head.

Targets can be represented in the exact same way: as a sphere with some distinct color. However, both target and end effector must be *visible* for both cameras. This puts a rather strong constraint on the target generation and on the task space overall, as only a fraction of the full kinematic mapping can be learned this way. The problem is circumvented by creating targets close to the end effector location; close enough to be visible, but far enough so that some effort is necessary to reach them.

---

[2] The camera parameters for the iCub are as follows: $f_x = 335$, $f_y = 250$, $c_x = 160$, $c_y = 120$. The cameras are $T_X = 68\,\text{mm}$ apart.

9.2 Learning to Control Arm and Head

The workflow of the framework in Section 8.2 consisted of six steps: sensor measurements, computing derivatives, model update, computation of the current Jacobian, inverse kinematics, and execution of the control commands (cf. Figure 8.2). The model was mapping arm joint velocities onto end effector motion. Those steps are now put into the context of the iCub including vision and object tracking.

Three different sensors have to be queried now: the arm joints as before and additionally head joints and the stereo cameras. Image processing reveals the pixel location of the end effector and the current target and reprojection yields the respective task space locations in the head reference frame. Since the three mentioned sensors send their measurements independently over the network, an alignment procedure is added to reduce noise in time. The relevant results are arm joint angles θ_a, head joint angles θ_h, the end effector location ξ and target location ξ^*. Derivatives are computed with respect to time as before.

The next step is learning of a kinematic model that maps arm *and* head joint velocities onto task space motion, that is, velocity $\dot{\xi}$ of the end effector in the head reference frame:

$$\dot{\xi} = J(\theta_{ah})\dot{\theta}_{ah}\,, \tag{9.3}$$

where $\theta_{ah} = (\theta_a, \theta_h)^T$ is the combination of all relevant joint angles and $J(\theta_{ah})$ is the Jacobian matrix for the current posture. While a non-zero head joint velocity does not change the location of the end effector in world coordinates, it does rotate the head's coordinate system and thus the location in the head reference frame is changed.

Given $n = n_a + n_h$ joints, with n_a arm joints, n_h head joints, and a task space of dimension m, the resulting Jacobian is a $n \times m$ matrix. It can be split into the respective head and arm parts, where the first n_a columns define the Jacobian for the arm motion and the latter n_h columns represent the matrix with respect to head movement.

The Jacobian of the arm is inverted with the goal of reaching the given target as previously done in Section 7.4. Thus, the desired task space velocity vector is the same as in Equation (7.18)

$$\dot{\xi}_a^* = \frac{v}{\|\xi^* - \xi\|}\xi^* - \xi\,, \tag{9.4}$$

where v is the desired velocity and the direction is from current location ξ towards the target ξ^*. The velocity v can be computed by a standard velocity profile as depicted in Figure 7.7. In turn, J_a and $\dot{\xi}_a^*$ can be put into Equation (7.17) for inversion with a suitable constraint.

The head Jacobian, however, is inverted to track the target, that is, to keep the target in a centered position in the head reference frame. The desired task space velocity is given by

$$\dot{\boldsymbol{\xi}}_h^* = \frac{v}{\sqrt{\xi_1^2 + \xi_2^2}} \begin{pmatrix} -\xi_1 \\ -\xi_2 \\ 0 \end{pmatrix}. \tag{9.5}$$

The third coordinate is not used, because it refers to depth in the head reference frame.

The above method centers the *target*, but not the *end effector*. Tracking the end effector itself is also possible, but this creates a severe problem for the learner: When the head Jacobian is accurate, the tracking is accurate which means that there is very little x or y motion in the head reference frame. As a consequence, the arm Jacobian cannot be trained well and the resulting control performance is very sloppy. Actually both, end effector and target, must be visible, so eventually the best solution in terms of visibility would be to track an intermediate point between end effector and target. However, targets are generated close to the current end effector location and tracking those targets does not leave the end effector out of sight unless the arm Jacobian points in the wrong direction and the arm moves away from the target. In this case vision is lost and no learning is possible to resolve the problem. Thus, the arm and head are reset to their default posture by a pre-programmed controller to ensure vision of the end effector. With a well tuned learning algorithm and suitable targets, this does not happen.

To sum up, the Function Approximation (FA) algorithm learns a mapping from arm *plus* head joint velocities to task space velocities. The corresponding Jacobian is split into its arm related and its head related columns, $\boldsymbol{J}_a$ and $\boldsymbol{J}_h$, respectively. The inverse kinematics of $\boldsymbol{J}_a$ yield the arm joint velocities required to reach the target, while the inverse of $\boldsymbol{J}_h$ is used to compute suitable head joint velocities to track the target. Secondary constraints can be used for both inversions, however, the head tracking here only uses two joints to control two directions, which leaves no freedom for additional constraints.

9.3 Experimental Validation

The iCub experiments follow a similar pattern than the previous experiments on the purely kinematic simulation, while the joint dynamics are handled by iCub's internal PID controllers. The modified control loop is illustrated in Figure 9.4. Training and test mode alternate every 1000 iterations and the same performance measurements are collected as before. A total training time of 50 k iterations is sufficient.

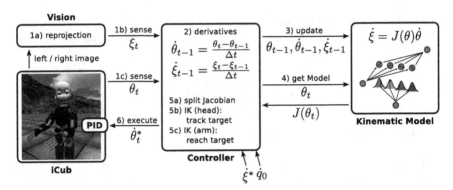

Figure 9.4: Mainly two steps are different compared to the previous, purely kinematic framework depicted in Figure 8.2: The first step, perception, now requires processing of the camera images and reprojection into a 3 D reference frame. The joint angles from head and arm are combined into a single vector $\boldsymbol{\theta}$. The fifth step, that is, inverse kinematics (IK), handles the head and arm kinematics separately. The resulting control signal can be directly sent to the iCub, whose internal PID controllers realize the desired head and arm velocities.

Training targets are generated uniformly randomly within a rectangular area in front of the robot. The rectangle is chosen such that significant head movement is required to "see" the corners, while extreme joint angles are not necessary. The learning is disabled during *test* mode, where a fixed set of nine targets form an asterisk shape of 10 cm radius within the rectangular training area [78, 24]. This allows to also visually inspect the resulting trajectories.

Where previously the majority of the control space was trained, the present setup is comparably simple. Therefore the maximum population size is set to 100 Receptive Fields (RFs). Other parameters are unchanged.

In order to track a target with the head, two degrees of freedom are required. Therefore the neck *pitch* and *yaw* joints are learned, while *roll* is disabled (fixed at zero angle). Furthermore four arm joints are enabled: three shoulder joints and the elbow flexion-extension (cf. Figure 8.3). This makes up for a total of six joints in the control space. As before the task space is three dimensional which results in a $\mathbb{R}^6 \mapsto \mathbb{R}^3$ kinematic mapping.

Twenty independent runs were conducted while minimum and maximum performance are recorded along with the median performance. The test performance shown in Figure 9.5 once again confirms that the forward kinematics can be learned quickly. The head kinematics are almost linear and learned fast enough that the iCub never lost sight of its target. Furthermore, all test targets are reached in every run and less then 100 RFs are sufficient to approximate the

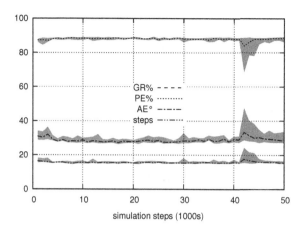

Figure 9.5: Learning is disabled during the test stage, where the robot's task is to draw an asterisk shape with its end effector.

kinematic mapping. After condensation, which begins at 40 k iterations, about 30 RFs remain in the population (minimum 19, maximum 36).

However, angular error and path efficiency are higher than in the purely kinematic experiments: The median of the final path efficiency is 88.5% (minimum 87.7%, maximum 89.0%) and the angular error is about 28.3° (minimum 27.3°, maximum 31.7°). Previously an angular error below 20° with a path efficiency greater 95% was achieved. This can be explained by several factors. First, the sensor measurement was exact before, while the iCub sensors are noisy in several aspects: Sensor values are sent over the network, which inherently produces a short temporal delay. Measurements from independent sources must be aligned in time, which requires interpolation and adds another small delay. Moreover, image processing produces another delay. Finally, the image reprojection is not exact, as depth information is based on pixel information and ultimately the precision depends on the camera resolution (here 320 × 240). Last but not least, the iCub simulator includes a physics engine and inertia naturally prevents an accurate asterisk shape, as the velocity points in the wrong direction on target change. Figure 9.6 depicts the trajectories during early and later test stages. Precision improves quickly and after less than 20 k learning steps, accuracy stabilizes.

In this chapter it was shown that the developed framework is robust enough to handle a more realistic scenario with noisy sensors. Here, the image streams of two cameras were processed to extract the location of the end effector. Further-

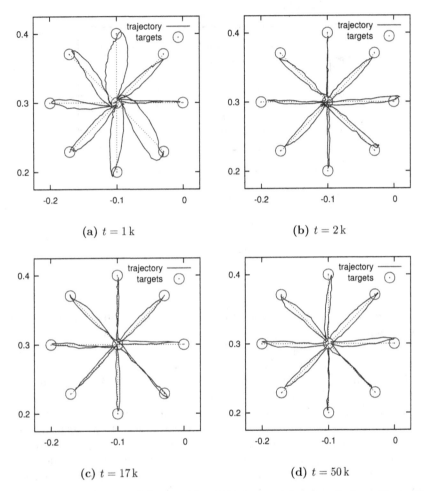

(a) $t = 1\,\mathrm{k}$ **(b)** $t = 2\,\mathrm{k}$

(c) $t = 17\,\mathrm{k}$ **(d)** $t = 50\,\mathrm{k}$

Figure 9.6: Trajectories of the asterisk task in horizontal and vertical *world* coordinates. The target distance from the robot (mostly related to depth in head coordinates) is constant for the targets. (a) The first test trial, where learning is disabled, starts after 1000 learning iterations. The asterisk trajectory is sloppy. (b) After 2000 iterations the movements become more precise. (c) The least angular error is found at around $17\,\mathrm{k}$ iterations. From then on, XCSF focuses on generalization and performance does not improve anymore. (d) After condensation, about 30 RFs remain in the population and the performance degrades only slightly.

more, not only the arm kinematics but also the head kinematics were learned and the resulting model was split for two different control strategies: Two neck joints were used to track a given target and four arm joints were successfully used to draw an asterisk shape of 10 cm radius.

10 Summary and Conclusion

The capabilities of the brain, be it human or animal, have always had a special appeal to researchers. How the learning of a body schema [45] and corresponding motor control could look like in the brain is the particular motivation of this thesis, approached from a computer science perspective. Different learning algorithms were evaluated and integrated into a control framework able to use a simulated, anthropomorphic arm to reach given target locations. The framework was not built to primarily achieve this task with maximum performance, but instead to model respective brain functionality.

Achieving this goal is not easy, though, and the present framework should be seen as a starting point. On the one hand, too little is known about the brain and on the other hand, too many details are known to be assembled easily. However, by carefully collecting requirements for a feasible learning structure, while seeking for a "working" solution results in an algorithmic finery that learns to move an anthropomorphic arm on its own.

This final chapter summarizes the achieved results, discusses the plausibility and limitations, and eventually opens up new research perspectives. The remaining structure partially resembles the three parts of the thesis. Section 10.1 reflects on Machine Learning (ML) and how a feasible and plausible learning algorithm could look like. A theoretical excursus is taken in Section 10.2, where the limits of such algorithms are revisited. The final framework is briefly explained in Section 10.3 and its cognitive plausibility is discussed. Section 10.4 outlines possible future directions.

10.1 Function Approximation in the Brain?

Certain ingredients are required when modeling *learning* of *motor skills* as discussed in the introductory Section 1.4. ML is the computational equivalent of human learning and it is particularly convenient to use Function Approximation (FA) techniques to learn the kinematic "function", which in turn can be used to control limb movements. Moreover, a plausible approach should build upon neuron like structures, where kinematics in particular are known to be represented in *directional population codes*. Finally, the prior knowledge of the model should be similar to the knowledge of a new born mammal. In other words, evolution may come up with a smart grounding but every individual must learn the motor skills on its own.

After basic and advanced FA methods were introduced, Chapter 3 outlined the elementary algorithmic features for learning a kinematic model. When directional population codes are modeled, there are three key features: clustering via kernels, the corresponding local models, and an inference procedure to merge individual models. A kernel represents the area of influence of a neuron, the local model represents a directional coding, and responses of multiple neurons are typically merged by a weighted sum. The simplified neuron view is justified by the fact that computers process data mostly serial while neurons work massively parallelized.

The outlined recipe provides a plausible basis for modeling brain functionality in line with neuro-scientific literature and, at the same time, it allows to build powerful learning algorithms without the need for immense computational resources. Three such algorithms were selected and evaluated throughout this work: Radial Basis Function Networks (RBFNs) are a particular type of Artificial Neural Network (ANN). The XCSF algorithm belongs to the family of Learning Classifier Systems (LCSs), which were envisioned as cognitive systems and, thus, naturally fit the task at hand. Finally, Locally Weighted Projection Regression (LWPR) is a more statistically grounded regression algorithm, which also contains the above named features.

10.2 Computational Demand of Neural Network Approximation

The neural network regression methods are – theoretically – able to reach any desired precision with sufficient resources as stated in the Cybenko theorem [23]. The problem is that "sufficient" could mean inappropriately large and therefore Chapters 5 and 6 take a closer look at the internal functioning and scalability of the XCSF algorithm. The general results carry over to other similar algorithms.

The main resource is the number of neurons (kernels), sometimes also called population size. How does the required number of neurons scale with the function complexity or input dimensionality? This question was addressed within a tractable scenario and basically confirms the curse of dimensionality: The population size scales exponentially with an increasing input dimension, because the dimension affects the volume to be covered with kernels in an exponential manner.

In the context of motor skill learning, however, it may not always be required to learn the whole space of movements equally well. Instead, emphasis should be put on generalization and specialization: Mammals are not equally proficient in their movements at any joint angle or at any side of their body (e.g. behind the back) [37]. Repeated movements should improve accuracy, while parts of the joint space that are rarely required need not to be learned precisely. This principle may lead to new learning algorithms that do not rely on minimizing an error, but instead continuously improve a clustering with limited resources (neurons), where regions of interest have more resources which indirectly results in a higher precision.

In terms of computational complexity the developed framework is *feasible* to be "implemented" in the brain, as the massive parallelization realized by asynchronously firing neurons allows to employ large populations of neurons for accurate approximation. However, this is not true for hardware with restricted computational resources, such as mobile robots. Few computational units are available to work in parallel and, thus, the number of Receptive Fields (RFs) is limited when *real time control* is desired. This essentially bounds the achievable model accuracy. It remains to be shown if sensor accuracy or computational demand is the limiting factor here – a highly hardware-related question.

10.3 Learning Motor Skills for Control

The learning algorithm is one ingredient and a control framework, that makes use of the learned model to control the body, is the other one. The basics of *control* were introduced in Chapter 7, while the follow up Chapter 8 integrated the *learning* of velocity kinematics.

In the context of an anthropomorphic arm with seven Degrees of Freedom (DoF) – just as the human arm – a simple control framework was built for reaching tasks. The general workflow of the control loop can be divided into three steps: sensor measurement, learning, and motor action. The motor commands are deduced from the learned kinematic model. Importantly, this works from scratch without any knowledge: initially, movements will be more or less random – because the model starts without knowledge –, but all tested learners quickly adapted their internal models to solve reaching tasks within few iterations. Longer training improved reaching accuracy and almost perfectly straight paths were taken towards given targets.

All tested learning algorithms (RBFN, XCSF, LWPR) have shown good performance on an exhaustive set of test targets distributed throughout the whole task space. A RBFN, the simplest algorithm of the three, does not modify the shape or size of its RFs, which on first sight may be a drawback. Indeed, XCSF was able to achieve slightly better performance with a lower number of neurons, since XCSF optimized the distribution, size, and shape of its RFs. However, RBFN's performance is very stable compared to the other algorithms and its simplicity is favorable over the complex algorithmic finery of XCSF and LWPR.

From this perspective, RBFN has the best "complexity to performance" ratio. This also reveals that learning the non-linear forward kinematics is more of a checkerboard problem, when locally linear models are employed. In this case, the advanced structuring capabilities of XCSF and LWPR can hardly improve the overall model accuracy.

The structural optimization of the neuron populations in XCSF and LWPR also comes at the cost of stability: a modified RF may produce inaccurate control signals until its local model is adapted. Nevertheless, such optimization can improve the control performance and reduce the number of neurons required to cover the joint space as shown for XCSF, which outperformed the RBFN in both population size and accuracy. This was shown for two different task space representations: a Cartesian end effector representation and an egocentric view, that represents the end effector by distance, vertical, and horizontal angle. The results confirmed that the framework generally does not depend on a particular representation, but on the other hand the difference in reaching accuracy has shown that some representations might be more suitable than others.

With respect to the three learning algorithms, the conclusion is that all of them are well-applicable to learning kinematics, while on the one hand RBFN shines through its simplicity and stable solution, but on the other hand XCSF achieved the best accuracy through smart structuring.

Moving from the theoretical kinematic simulator towards a more realistic scenario, experiments with the iCub robot were conducted in Chapter 9. The iCub simulation comes with a physics engine and also allows to use stereo cameras for detection of the end effector or targets instead of the exact simulation values used before, which are not available for humans, either. This leads to a visual servoing [21] framework, where vision is used to control robot movements. Furthermore, the noise inherent to vision puts the robustness of the framework to the test.

When target and end effector location are perceived by cameras, the head must be moved such that both, target and end effector, are visible. Thus, the kinematic model of the head was learned along with the arm model. For control, the model was separated, simply by splitting a matrix in two parts. The head related control procedure tracked the target, while the arm joints were used to reach for the target. The results are satisfactory and confirm the robustness and flexibility of the developed framework.

However, this method may not scale well: Reaching movements can involve a lot more joints than just arm and head, but also torso, legs, or the other arm as well. Combining all those joints into one single model brings a high dimensionality, which in turn demands for a large number of neurons. Using a uniform neuron distribution (the RBFN approach) results in an exponential number of neurons. Using the on-demand creation in XCSF or LWPR does not solve this issue directly, as both algorithms must first optimize the structure before it can be better then the uniform neuron distribution. Thus, the learning stage may require even more neurons and a considerable amount of optimization.

However, when not the full space of joint angles is required or at least less priority is given to rare joint configurations, learning algorithms may be able to handle the relevant areas. In high dimensional problems, such algorithms must start with a general solution (comparably few, large RFs) and iteratively refine the clustering by specializing in directions of large errors and eventually generalizing in directions, where the local model is sufficiently accurate. This refining process, however, is non trivial and currently no solution is applicable to learning on kinematic trajectories: LWPR's adaptation method is fragile in terms of parameter setup and does not approach an optimal clustering, unless almost spherical RFs are sufficient [87]. "Guided" XCSF on the other hand cannot handle trajectory data well, but requires uniformly distributed inputs.

10.3.1 Retrospective: Is it Cognitive and Plausible?

Several features known to be found in the brain were implemented within the developed framework: A neural network structure is used to learn a *body schema*, that is, the relation between joint angles and the end effector location, which defines the reachable (peripersonal) space [68, 45]. A population of RFs represents a *directional population code* of neurons and learning the velocity kinematics is in line with findings in the literature [1, 65, 68, 76, 11, 30]. The use of *forward-inverse models* allows to exploit the full flexibility of the body, especially in constrained circumstances [101, 39, 42].

Furthermore, the kinematic mapping is learned from scratch, learned *for action*. The motivation to act, however, is an external developer imposed motivation, but we have to start somewhere. Learning from scratch and using only this learned knowledge for control is important, as it closes a circle: sense, learn, and act. It is not necessary to initialize the model with external knowledge.

However, some algorithmic details are questionable. It is unlikely that the brain runs a Singular Value Decomposition (SVD) to invert the kinematic model. On the other hand, it was shown that comparably few neurons can implement a complex algorithm such as a Kalman filter [95]. Thus, not all details may be 100% plausible, but the general idea remains valid. Additionally, computers work on the bit level and mostly serial, while the brain runs millions of parallelized neurons – thus the simplifications seem reasonable.

Personally, I feel that something more elaborate than RBFN is required to include long term adaptation to the environment, while the complex algorithmic details of XCSF and LWPR are difficult to justify. Overall, the approach lives more on the bottom-up side where the connection to neuroscience is well accomplished with some simplifications that have to be made. However, towards the psychological research side (top-down view on the brain) there is a large gap. In its present state the model cannot explain more interesting psychological phenomena – apart from plain learning – and future work should extend into this research direction.

10.3.2 On Optimization and Inverse Control

The considered FA algorithms strive to minimize their approximation error, which is less useful for learning kinematics for the following reason. The largest errors originate from shoulder movements, as the shoulder joint is the farthest away from the end effector. On the contrary, a 1° error in the wrist joint has little effect on the reaching error. Thus, most resources are assigned to the shoulder dimensions, while wrist joints often receive a comparably coarse clustering. On the contrary, humans instead show high dexterity in wrist and fingers. Again, this suggests that optimization based on a simple prediction error does not happen in the brain.

Furthermore, a generated movement plan may not use the wrist joints at all, because shoulder joint movements reach a given target faster, that is, with less overall joint angle displacement. This is due to the inverse kinematics, basically a matrix inversion, which by default favors a minimum norm solution. Minimum norm means that the sum of joint velocities is minimized by default and only the secondary constraints may request extra effort from the nullspace. Eventually, a weighted matrix inversion results in more realistic joint usage. Instead of minimizing the absolute sum of joint velocities, the required (muscle) energy could be considered.

10.4 Outlook

Despite of convincing results and its plausibility, the developed framework lacks an important feature: sophisticated planning. In this thesis only simple reaching tasks where considered, where a straight path is available, which is not necessarily the case: Obstacles could require advanced trajectories. Therefore, a planning algorithm has to predict a path in the task space, to reach a given target without collisions. Effective collision prevention will require models for each joint, not only the end effector, which may be an extension to this work. Furthermore, forward predictions of the kinematic model could be used to evaluate movement alternatives.

One might argue that it is critical to include dynamics into the learned models. However, muscles are very different from servo motors: A servo motor spins in two directions, or put differently, is able to accelerate *and* decelerate. Muscles, on the other hand, work in pairs or larger groups to jointly control joints, because they can only contract (strong force) or relax (negligible force). Morphological intelligence [64] plays a large role to ease control and it is non trivial to model muscle groups accurately. Therefore the explanatory power of dynamic models for servo motors is questionable in the light of cognitive science.

Another challenging research topic is how exactly the task space is built. A green sphere was used in Chapter 9 to detect the end effector, which simplified the definition of the task space. Using *pure* vision to learn the effects of limb movements, the respective reachable space, and thus the task space is challenging, especially in a realistic scenario. Instead, learning the task space should include *other* sensory cues such as touch and proprioception to complement the visual feedback. Such a model would define the task space in multiple "input" modalities, however, with a common "output" coordinate system. In other words, one model would map proprioceptive feedback to resulting task space position, another model would map visual cues to the task space, and a third one could be based on touch. Such a multimodal representation would be robust to sensor issues (e.g. occluded vision) that would disrupt control otherwise. This also starts to connect to higher level cognitive phenomena such as the rubber hand illusion [9], where a rubber hand is *perceived* as one's own hand, because two sensory cues (vision and touch) are able to override proprioceptive feedback.

Thinking about controlling a whole body with numerous joints and various sensors, algorithms have to scale up for higher-dimensional tasks while smart divide and conquer strategies must disassemble the large mappings into independent groups that can be recombined on a higher, simple level. Which body parts or modalities must be grouped depends on the task space, as shown with the iCub, where both head and arm movements affected the task space location in the *head* coordinate system. In a *torso* centered coordinate system instead both mappings could be learned separately, eventually with fewer resources and less learning time. Thus, the choice of the task space is a crucial ingredient of modeling kinematic learning.

Concerning the learning algorithms and higher dimensional problems, a start has been made with guided XCSF in Section 6.2, which however is based on the rather complex algorithmic finery of a LCS. Combining the simplicity of RBFN with a similar idea to "guide" the shape, size, and location of RFs may result in a simple, yet powerful learning tool to model brain functionality in future works.

Bibliography

[1] R. Ajemian, D. Bullock, and S. Grossberg. Kinematic coordinates in which motor cortical cells encode movement direction. *Journal of Neurophysiology*, 84(5):2191–2203, 2000. (citations on pages 6 and 141)

[2] E. Anderson, Z. Bai, C. Bischof, S. Blackford, J. Demmel, J. Dongarra, J. Du Croz, A. Greenbaum, S. Hammarling, A. McKenney, and D. Sorensen. *LAPACK Users' Guide*. Society for Industrial and Applied Mathematics, third edition, 1999. (citation on page 97)

[3] K. J. Åström and B. Wittenmark. *Adaptive Control*. Addison-Wesley, second edition, 1995. (citations on pages 15, 16, and 17)

[4] J. Baillieul, J. Hollerbach, and R. Brockett. Programming and control of kinematically redundant manipulators. In *Decision and Control, The 23rd IEEE Conference on*, volume 23, pages 768–774, 1984. (citation on page 91)

[5] D. R. Baker and I. Wampler, Charles W. On the inverse kinematics of redundant manipulators. *The International Journal of Robotics Research*, 7(2):3–21, 1988. (citation on page 91)

[6] R. H. Bartels, J. C. Beatty, and B. A. Barsky. *An Introduction to Splines for Use in Computer Graphics and Geometric Modelling*, chapter 3, pages 9–17. The Morgan Kaufmann Series in Computer Graphics. Morgan Kaufmann, 1995. (citation on page 18)

[7] A. Ben-Israel and T. N. Greville. *Generalized Inverses: Theory and Applications*. Springer, 2003. (citation on page 92)

[8] N. E. Berthier, M. T. Rosenstein, and A. G. Barto. Approximate optimal control as a model for motor learning. *Psychological Review*, 112(2):329–346, 2005. (citation on page 5)

[9] M. Botvinick and J. Cohen. Rubber hands 'feel' touch that eyes see. *Nature*, 391:756, 1998. (citation on page 143)

[10] G. Bradski. The OpenCV Library. *Dr. Dobb's Journal of Software Tools*, 2000. (citation on page 97)

[11] C. A. Buneo, M. R. Jarvis, A. P. Batista, and R. A. Andersen. Direct visuomotor transformations for reaching. *Nature*, 416(6881):632–636, 2002. (citations on pages 6 and 141)

[12] C. J. C. Burges. A tutorial on support vector machines for pattern recognition. *Data Mining and Knowledge Discovery*, 2(2):121–167, 1998. (citation on page 30)

[13] M. V. Butz. *Rule-Based Evolutionary Online Learning Systems: A Principal Approach to LCS Analysis and Design*. Springer, 2006. (citations on pages 57 and 58)

[14] M. V. Butz and O. Herbort. Context-dependent predictions and cognitive arm control with XCSF. In *GECCO '08: Proceedings of the 10th Annual Conference on Genetic and Evolutionary Computation*, pages 1357–1364. ACM, 2008. (citations on pages 110 and 115)

[15] M. V. Butz, O. Herbort, and J. Hoffmann. Exploiting redundancy for flexible behavior: Unsupervised learning in a modular sensorimotor control architecture. *Psychological Review*, 114:1015–1046, 2007. (citation on page 5)

[16] M. V. Butz, T. Kovacs, P. L. Lanzi, and S. W. Wilson. How XCS evolves accurate classifiers. In *Proceedings of the Genetic and Evolutionary Computation Conference (GECCO-2001)*, pages 927–934, 2001. (citations on pages 57 and 60)

[17] M. V. Butz, T. Kovacs, P. L. Lanzi, and S. W. Wilson. Toward a theory of generalization and learning in XCS. *IEEE Transactions on Evolutionary Computation*, 8:28–46, 2004. (citation on page 57)

[18] M. V. Butz, P. L. Lanzi, and S. W. Wilson. Function approximation with XCS: Hyperellipsoidal conditions, recursive least squares, and compaction. *IEEE Transactions on Evolutionary Computation*, 12:355–376, 2008. (citations on pages 53 and 60)

[19] M. V. Butz, G. K. Pedersen, and P. O. Stalph. Learning sensorimotor control structures with XCSF: Redundancy exploitation and dynamic control. In *GECCO '09: Proceedings of the 11th Annual Conference on Genetic and Evolutionary Computation*, pages 1171–1178, 2009. (citations on pages 110 and 115)

[20] S. Chiaverini, O. Egeland, and R. Kanestrom. Achieving user-defined accuracy with damped least-squares inverse kinematics. In *Fifth Interna-*

tional Conference on Advanced Robotics, volume 1, pages 672–677, june 1991. (citation on page 95)

[21] P. I. Corke. *Visual Control of Robots: High-Performance visual servoing*, volume 2 of *Mechatronics*. Research Studies Press (John Wiley), 1996. (citations on pages 125 and 140)

[22] J. J. Craig. *Introduction to Robotics: Mechanics and Control*. Addison-Wesley Longman Publishing Co., Inc., Boston, MA, USA, 1989. (citation on page 91)

[23] G. Cybenko. Approximation by superpositions of a sigmoidal function. *Mathematics of Control, Signals, and Systems*, 2:303–314, 1989. (citations on pages 21, 37, 57, and 138)

[24] A. Droniou, S. Ivaldi, V. Padois, and O. Sigaud. Autonomous online learning of velocity kinematics on the iCub: a comparative study. In *Proceedings IEEE/RSJ International Conference on Intelligent Robots and Systems*, pages 5377–5382, 2012. (citations on pages 110, 115, 125, and 132)

[25] A. Droniou, S. Ivaldi, P. O. Stalph, M. V. Butz, and O. Sigaud. Learning velocity kinematics: Experimental comparison of on-line regression algorithms. In *Proceedings of Robotica 2012*, pages 7–13, 2012. (citations on pages 110 and 115)

[26] J. Duchon. Splines minimizing rotation-invariant semi-norms in sobolev spaces. In W. Schempp and K. Zeller, editors, *Constructive Theory of Functions of Several Variables*, volume 571 of *Lecture Notes in Mathematics*, pages 85–100. Springer Berlin Heidelberg, 1977. (citation on page 30)

[27] J. W. Eaton. *GNU Octave Manual*. Network Theory Limited, 2002. (citation on page 97)

[28] G. E. Forsythe, M. A. Malcolm, and C. B. Moler. *Computer Methods for Mathematical Computations*, chapter 9. Prentice Hall, 1977. (citation on page 93)

[29] C. F. Gauss. *Theoria Motus Corporum Coelestium in sectionibus conicis solem ambientium*. Göttingen, 1809. (citation on page 13)

[30] A. P. Georgopoulos. Current issues in directional motor control. *Trends in Neurosciences*, 18(11):506–510, 1995. (citations on pages 6 and 141)

[31] A. P. Georgopoulos, M. Taira, and A. Lukashin. Cognitive neurophysiology of the motor cortex. *Science*, 260(5104):47–52, Apr 1993. (citation on page 6)

[32] W. Gerstner. Spiking neurons. In W. Maass and C. M. Bishop, editors, *Pulsed Neural Networks*, chapter 1, pages 3–54. MIT Press, 2001. (citation on page 21)

[33] G. Golub and W. Kahan. Calculating the singular values and pseudo-inverse of a matrix. *Journal of the Society for Industrial and Applied Mathematics*, 2(2):205–224, 1965. (citation on page 92)

[34] G. Golub and C. Reinsch. Singular value decomposition and least squares solutions. *Numerische Mathematik*, 14:403–420, 1970. (citations on pages 75 and 93)

[35] G. H. Golub and C. F. Van Loan. *Matrix Computations*, volume 3. Johns Hopkins University Press, 1996. (citation on page 75)

[36] A. Graves, M. Liwicki, S. Fernandez, R. Bertolami, H. Bunke, and J. Schmidhuber. A novel connectionist system for unconstrained handwriting recognition. *Pattern Analysis and Machine Intelligence, IEEE Transactions on*, 31(5):855–868, 2009. (citation on page 21)

[37] M. Graziano. The organization of behavioral repertoire in motor cortex. *Annual Review of Neuroscience*, 29:105–134, 2006. (citations on pages 108 and 139)

[38] W. E. Hart, N. Krasnogor, and J. E. Smith. *Recent Advances in Memetic Algorithms*, volume 166 of *Studies in Fuzziness and Soft Computing*. Springer, 2005. (citation on page 73)

[39] M. Haruno, D. M. Wolpert, and M. Kawato. MOSAIC model for sensorimotor learning and control. *Neural Computation*, 13(10):2201–2220, 2001. (citations on pages 5 and 141)

[40] S. Haykin. *Neural Networks - A Comprehensive Foundation*, chapter 5, pages 278–339. Prentice Hall, 2nd edition, 1999. (citations on pages 21 and 30)

[41] S. O. Haykin. *Adaptive Filter Theory*. Prentice Hall, 4th edition, 2001. (citations on pages 15, 16, and 17)

[42] O. Herbort, M. V. Butz, and G. Pedersen. The SURE_REACH model for motor learning and control of a redundant arm: From modeling human behavior to applications in robotics. In O. Sigaud and J. Peters, editors, *From Motor Learning to Interaction Learning in Robots*, volume 264 of *Studies in Computational Intelligence*, pages 85–106. Springer, 2010. (citation on page 141)

[43] J. H. Holland. *Adaptation in natural and artificial systems: An intro-ductory analysis with applications to biology, control, and artificial intelligence.* The MIT Press, Cambridge, Massachusetts, 1992. (citation on page 41)

[44] J. H. Holland and J. S. Reitman. Cognitive systems based on adaptive algorithms. *SIGART Bull.*, 63(63):49–49, 1977. (citation on page 41)

[45] N. P. Holmes and C. Spence. The body schema and multisensory representation(s) of peripersonal space. *Cognitive Processing*, 5:94–105, 2004. (citations on pages 137 and 141)

[46] F. C. Hoppensteadt and E. M. Izhikevich. *Weakly Connected Neural Networks*, volume 126 of *Applied Mathematical Sciences*, chapter 1, pages 1–22. Springer, 1997. (citation on page 5)

[47] J. Hurst and L. Bull. A self-adaptive neural learning classifier system with constructivism for mobile robot control. In X. Yao, E. Burke, J. Lozano, J. Smith, J. Merelo-Guervós, J. Bullinaria, J. Rowe, P. Tino, A. Kabán, and H.-P. Schwefel, editors, *Parallel Problem Solving from Nature - PPSN VIII*, volume 3242 of *Lecture Notes in Computer Science*, pages 942–951. Springer, 2004. (citation on page 34)

[48] M. I. Jordan and D. E. Rumelhart. Forward models: Supervised learning with a distal teacher. *Cognitive Science*, 16(3):307–354, 1992. (citation on page 5)

[49] J. H. Kaas. Plasticity of sensory and motor maps in adult mammals. *Annual Review of Neuroscience*, 14(1):137–167, 1991. (citation on page 5)

[50] J. H. Kaas. The reorganization of sensory and motor maps after injury in adult mammals. In M. S. Gazzaniga, editor, *The new Cognitive Neurosciences*, pages 223–235. MIT Press, second edition, 1999. (citation on page 5)

[51] R. Kalman. A new approach to linear filtering and prediction problems. *Journal of Basic Engineering*, 82(D):35–45, 1960. (citation on page 17)

[52] M. Kawato, K. Furukawa, and R. Suzuki. A hierarchical neural-network model for control and learning of voluntary movement. *Biological Cybernetics*, 57(3):169–185, 1987. (citation on page 5)

[53] J. F. Kenney and E. S. Keeping. *Mathematics of Statistics, Pt. 2*, chapter 8, pages 199–237. Princeton, N.J: Van Nostrand, 2nd edition, 1951. (citation on page 14)

[54] S. Klanke and S. Vijayakumar. LWPR implementation in C (v. 1.2.4). http://sourceforge.net/projects/lwpr/files/lwpr-1.2.4/. [Online, Accessed: 04-March-2012]. (citation on page 102)

[55] P. L. Lanzi, D. Loiacono, S. W. Wilson, and D. E. Goldberg. Extending XCSF beyond linear approximation. In *GECCO '05: Proceedings of the 2005 Conference on Genetic and Evolutionary Computation*, pages 1827–1834, 2005. (citation on page 36)

[56] P. L. Lanzi, D. Loiacono, S. W. Wilson, and D. E. Goldberg. Prediction update algorithms for XCSF: RLS, kalman filter, and gain adaptation. In *GECCO '06: Proceedings of the 8th Annual Conference on Genetic and Evolutionary Computation*, pages 1505–1512. ACM, 2006. (citation on page 43)

[57] A.-M. Legendre. *Nouvelles méthodes pour la détermination des orbites des comètes*. Paris, 1805. (citation on page 13)

[58] A. Liegeois. Automatic supervisory control of the configuration and behavior of multibody mechnisms. *IEEE Transactions on Systems, Man, and Cybernetics*, 7(12):868–871, 1977. (citation on page 92)

[59] D. Loiacono and P. Lanzi. Evolving neural networks for classifier prediction with XCSF. In *Proceedings of the Workshop on Evolutionary Computation (ECAI'06)*, pages 36–40, 2006. (citation on page 34)

[60] A. Maravita, C. Spence, and J. Driver. Multisensory integration and the body schema: Close to hand and within reach. *Current Biology*, 13(13):R531–R539, 2003. (citation on page 6)

[61] W. S. McCulloch and W. H. Pitts. A logical calculus of the ideas immanent in nervous activity. *Bulletin of Mathematical Biophysics*, 5:115–133, 1943. (citations on pages 20 and 21)

[62] P. Moscato, R. Berretta, and C. Cotta. *Memetic Algorithms*. John Wiley & Sons, Inc., 2010. (citation on page 73)

[63] Y. Nakamura and H. Hanafusa. Inverse kinematic solutions with singularity robustness for robot manipulator control. *Journal of Dynamic Systems, Measurement, and Control*, 108(3):163–171, 1986. (citation on page 94)

[64] R. Pfeifer and J. Bongard. *How the Body Shapes the Way We Think*. MIT Press, 2006. (citation on page 143)

[65] A. Pouget, P. Dayan, and R. Zemel. Information processing with population codes. *Nature Reviews Neuroscience*, 1:125–132, 2000. (citations on pages 6 and 141)

[66] W. H. Press, S. A. Teukolsky, W. T. Vetterling, and B. P. Flannery. *Numerical Recipes*, chapter 2, pages 37–109. Cambridge University Press, 2007. (citations on pages 75, 92, and 93)

[67] C. E. Rasmussen and C. K. I. Williams. *Gaussian Processes for Machine Learning*. MIT Press, 2nd edition, 2006. (citations on pages 19, 20, 30, and 31)

[68] G. Rizzolatti, L. Fadiga, L. Fogassi, and V. Gallese. The space around us. *Science*, 277(5323):190–191, 1997. (citations on pages 6 and 141)

[69] RobotCub Consortium. An open source cognitive humanoid robotic platform. http://www.icub.org/. [Online, Accessed: 17-September-2012]. (citations on pages 1, 4, and 125)

[70] M. Rolf, J. J. Steil, and M. Gienger. Goal babbling permits direct learning of inverse kinematics. *Autonomous Mental Development, IEEE Transactions on*, 2(3):216–229, 2010. (citation on page 112)

[71] D. A. Rosenbaum, R. G. J. Meulenbroek, J. Vaughan, and C. Jansen. Posture-based motion planning: Applications to grasping. *Psychological Review*, 108(4):709–734, 2001. (citation on page 5)

[72] M. Rosenblatt. Remarks on some nonparametric estimates of a density function. *Annals of Mathematical Statistics*, 27(3):832–837, 1956. (citations on pages 30 and 31)

[73] H. Sahbi. Kernel PCA for similarity invariant shape recognition. *Neurocomputing*, 70:3034–3045, 2007. (citation on page 31)

[74] C. Salaün, V. Padois, and O. Sigaud. Control of redundant robots using learned models: An operational space control approach. In *Proceedings of the IEEE/RSJ International Conference on Intelligent Robots and Systems*, pages 878–885, 2009. (citation on page 110)

[75] S. Schaal and C. G. Atkeson. Constructive incremental learning from only local information. *Neural Computation*, 10(8):2047–2084, 1998. (citations on pages 25 and 26)

[76] A. B. Schwartz, D. W. Moran, and G. A. Reina. Differential representation of perception and action in the frontal cortex. *Science*, 303(5656):380–383, 2004. (citations on pages 6 and 141)

[77] R. Shadmehr, M. A. Smith, and J. W. Krakauer. Error correction, sensory prediction, and adaptation in motor control. *Annu Rev Neurosci*, 33:89–108, 2010. (citation on page 5)

[78] G. Sicard, C. Salaün, S. Ivaldi, V. Padois, and O. Sigaud. Learning the velocity kinematics of iCub for model-based control: XCSF versus LWPR. In *Humanoid Robots (Humanoids), 11th IEEE-RAS International Conference on*, pages 570–575, 2011. (citations on pages 110, 115, 125, and 132)

[79] B. Siciliano and O. Khatib. *Springer Handbook of Robotics*. Springer, 2007. (citations on pages 91, 97, and 126)

[80] P. O. Stalph and M. V. Butz. Documentation of JavaXCSF. Technical Report Y2009N001, COBOSLAB, Department of Psychology III, University of Würzburg, Röntgenring 11, 97070 Würzburg, Germany, October 2009. (citation on page 102)

[81] P. O. Stalph and M. V. Butz. How fitness estimates interact with reproduction rates: Towards variable offspring set sizes in XCSF. In *Learning Classifier Systems*, volume 6471 of *LNCS*, pages 47–56. Springer, 2010. (citation on page 46)

[82] P. O. Stalph and M. V. Butz. Guided evolution in XCSF. In *GECCO '12: Proceedings of the 14th Annual Conference on Genetic and Evolutionary Computation*, pages 911–918, 2012. (citations on pages 47 and 72)

[83] P. O. Stalph and M. V. Butz. Learning local linear jacobians for flexible and adaptive robot arm control. *Genetic Programming and Evolvable Machines*, 13(2):137–157, 2012. (citations on pages 102, 110, 115, 116, 118, and 120)

[84] P. O. Stalph, M. V. Butz, D. E. Goldberg, and X. Llorà. On the scalability of XCS(F). In *GECCO '09: Proceedings of the 11th Annual Conference on Genetic and Evolutionary Computation*, pages 1315–1322, 2009. (citations on pages 110 and 115)

[85] P. O. Stalph, M. V. Butz, and G. K. Pedersen. Controlling a four degree of freedom arm in 3D using the XCSF learning classifier system. In *KI 2009: Advances in Artificial Intelligence*, volume 5803 of *LNCS*, pages 193–200, 2009. (citation on page 116)

[86] P. O. Stalph, X. Llorà, D. E. Goldberg, and M. V. Butz. Resource management and scalability of the XCSF learning classifier system. *Theoretical Computer Science*, 425:126–141, March 2012. (citations on pages 64 and 69)

[87] P. O. Stalph, J. Rubinsztajn, O. Sigaud, and M. V. Butz. Function approximation with LWPR and XCSF: a comparative study. *Evolutionary Intelligence*, 5(2):103–116, 2012. (citations on pages 114 and 141)

[88] C. Stone and L. Bull. An analysis of continuous-valued representations for learning classifier systems. In L. Bull and T. Kovacs, editors, *Foundations of Learning Classifier Systems*, Studies in Fuzziness and Soft Computing, pages 127–175. Springer, 2005. (citation on page 69)

[89] R. J. Urbanowicz and J. H. Moore. Learning classifier systems: A complete introduction, review, and roadmap. *Journal of Artificial Evolution and Applications*, 2009:1–25, 2009. (citation on page 41)

[90] G. Venturini. Adaptation in dynamic environments through a minimal probability of exploration. In *Proceedings of the Third International Conference on Simulation of Adaptive Behavior: From Animals to Animats 3 (SAB94)*, pages 371–379, 1994. (citation on page 44)

[91] S. Vijayakumar, A. D'Souza, and S. Schaal. Incremental online learning in high dimensions. *Neural Computation*, 17(12):2602–2634, 2005. (citations on pages 25, 26, and 77)

[92] S. Vijayakumar and S. Schaal. Locally weighted projection regression: An O(n) algorithm for incremental real time learning in high dimensional space. In *ICML '00: Proceedings of the Seventeenth International Conference on Machine Learning*, pages 1079–1086, 2000. (citation on page 30)

[93] C. Wampler. Manipulator inverse kinematic solutions based on vector formulations and damped least-squares methods. *Systems, Man and Cybernetics, IEEE Transactions on*, 16(1):93–101, 1986. (citation on page 94)

[94] D. E. Whitney. Resolved motion rate control of manipulators and human prostheses. *IEEE Transactions on Man-Machine Systems*, 10(2):47–53, 1969. (citation on page 5)

[95] R. Wilson and L. Finkel. A neural implementation of the Kalman filter. In *Advances in Neural Information Processing Systems*, volume 22, pages 2062–2070, 2009. (citations on pages 27 and 141)

[96] S. W. Wilson. Classifier fitness based on accuracy. *Evolutionary Computation*, 3(2):149–175, 1995. (citation on page 41)

[97] S. W. Wilson. Generalization in the XCS classifier system. *Genetic Programming 1998: Proceedings of the Third Annual Conference*, pages 665–674, 1998. (citations on pages 50, 52, and 60)

[98] S. W. Wilson. Function approximation with a classifier system. In *GECCO '01: Proceedings of the 2001 conference on Genetic and evolutionary computation*, pages 974–981. Morgan Kaufmann, 2001. (citation on page 26)

[99] S. W. Wilson. Classifiers that approximate functions. *Natural Computing*, 1:211–234, 2002. (citations on pages 26, 30, and 41)

[100] S. W. Wilson. Classifier conditions using gene expression programming. In *Learning Classifier Systems, Revised Selected Papers of IWLCS 2006-2007*, Lecture Notes in Artificial Intelligence, pages 206–217. Springer, 2008. (citation on page 34)

[101] D. M. Wolpert and M. Kawato. Multiple paired forward and inverse models for motor control. *Neural Networks*, 11:1317–1329, 1998. (citations on pages 5 and 141)

A A one-dimensional Toy Problem for Regression

A one-dimensional function is given as

$$f(x) = \sin(10\exp(-2x))\exp(-2x).$$

Ten samples are drawn uniformly random from the $[0,1]$ interval in Table A.1.

random points x	function value $f(x)$
0.64854884487441633350	0.10853658816403833696
0.46336252937406537064	-0.28858051577896974012
0.73628955961790826136	0.17202998459673039194
0.80803857539597766045	0.18173443306623248367
0.23071993163853877376	0.01296468819232347220
0.70513016144291512802	0.15738890146836519377
0.05044709616382335886	0.33908984659479130421
0.92889187292092654194	0.15600923143041300085
0.33433027130954924161	-0.46960934505296456386
0.41981261635181737723	-0.39883810785647795237

Table A.1